A2-Level
Biology

A2 Biology is seriously tricky — no question about that.
To do well, you're going to need to revise properly and practise hard.

This book has thorough notes on all the theory you need,
and it's got practice questions... lots of them.
For every topic there are warm-up and exam-style questions.

And of course, we've done our best to make the whole thing vaguely entertaining for you.

Complete Revision and Practice
Exam Board: Edexcel

Published by CGP

Editors:
Ellen Bowness, Katie Braid, Joe Brazier, Charlotte Burrows, Katherine Craig,
Rosie Gillham, Murray Hamilton, Jane Towle.

Contributors:
Gloria Barnett, Jessica Egan, Mark Ellingham, James Foster, Julian Hardwick, Derek Harvey,
Adrian Schmit, Sophie Watkins.

Proofreader:
Glenn Rogers.

ISBN: 978 1 84762 264 8

With thanks to Laura Stoney for the copyright research.

*Data used to construct the graph of temperature change over the last 1000 years on page 20 reproduced with kind
permission from Climate Change 2001: The Scientific Basis, Contribution of Working Group I to the Third Assessment
Report of the Intergovernmental Panel on Climate Change, SPM Figure 1. Cambridge University Press.*

*Data used to construct the graph of methane concentration on page 22 © CSIRO Marine and Atmospheric Research,
reproduced with permission from www.csiro.au.*

*Data used to construct the graph of CO_2 concentration on page 22 reproduced with kind permission from
U.S. Global Change Research Program, http://www.usgcrp.gov/usgcrp/nacc/background/scenarios/images/co2hm.gif.*

*Graph of emissions scenarios on page 27 modified and based on Special Report of Working Group III
of the Intergovernmental Panel on Climate Change on Emissions Scenarios (IPCC 2000).*

Photographs on pages 85 and 86 reproduced with kind permission from Science Photo Library.

*Data on exercise and obesity on page 68 adapted by permission from Macmillan Publishers Ltd:
M.Á. Martínez-González, J.A. Martínez, F.B. Hu, M.J. Gibney, J. Kearney. Physical inactivity, sendentary lifestyle
and obesity in the European Union. International Journal of Obesity; 23: 1192-1201, copyright 1999.*

*Data used to construct the graph on exercise and coronary heart disease on page 68 from S.G. Wannamethee, A.G. Shaper,
K.G. Alberti. Physical activity, metabolic factors, and the incidence of coronary heart disease and type 2 diabetes.
Archives of Internal Medicine, 2000; 160:2108-2116. Copyright © 2000 American Medical Association. All rights reserved.*

*Data used to construct the graph on exercise and type 2 diabetes on page 68 from S.G. Wannamethee, A.G. Shaper,
K.G. Alberti. Physical activity, metabolic factors, and the incidence of coronary heart disease and type 2 diabetes.
Archives of Internal Medicine, 2000; 160:2108-2116. Copyright © 2000 American Medical Association. All rights reserved.*

*Data on exercise and osteoarthritis on page 69 reproduced from U.M. Kujala, J. Kapiro, S. Sarna.
Osteoarthritis of weight bearing joints of lower limbs in former elite male athletes.
BMJ 2004; 308:231-234. Data reproduced with permission from BMJ Publishing Group Ltd.*

*Data on exercise and the immune system on page 69 from L. Spence, W.J. Brown, D.B. Pyne, M.D. Nissen,
T.P. Sloots, J.G. McCormack, A.S. Locke, P.A. Fricker. Incidence, etiology and symptomatology
of upper respiratory illness in elite athletes. Med Sci Sports Exerc 2007; 39:577-586.*

*Data in the exam question on page 69 from T.Y. Li, J.S. Rana, J.E. Manson, W.C. Willett, M.J. Stampfer,
G.A. Colditz, K.M. Rexrode, F.B. Hu. Obesity as compared with physical activity in predicting risk
of coronary heart disease in women. Circulation 2006; 113:499-506.*

Groovy website: www.cgpbooks.co.uk
Jolly bits of clipart from CorelDRAW®
Printed by Elanders Ltd, Newcastle upon Tyne.

Based on the classic CGP style created by Richard Parsons.

Contents

The Scientific Process

This stuff may look similar to what you learnt at AS, but that's because you need to understand How Science Works for A2 as well. 'How Science Works' is all about the scientific process — how we develop and test scientific ideas. It's what scientists do all day, every day (well, except at coffee time — never come between a scientist and their coffee).

Scientists Come Up with **Theories** — Then **Test Them**...

Science tries to explain **how** and **why** things happen — it **answers questions**. It's all about seeking and gaining **knowledge** about the world around us. Scientists do this by **asking** questions and **suggesting** answers and then **testing** them, to see if they're correct — this is the **scientific process**.

1) **Ask** a question — make an **observation** and ask **why or how** it happens. E.g. why do plants grow faster in glasshouses than outside?

2) **Suggest** an answer, or part of an answer, by forming a **theory** (a possible **explanation** of the observations), e.g. glasshouses are warmer than outside and plants grow faster when it's warmer because the rate of photosynthesis is higher. (Scientists also sometimes form a **model** too — a **simplified picture** of what's physically going on.)

3) Make a **prediction** or **hypothesis** — a **specific testable statement**, based on the theory, about what will happen in a test situation. E.g. the rate of photosynthesis will be faster at 20 °C than at 10 °C.

4) Carry out a **test** — to provide **evidence** that will support the prediction (or help to disprove it). E.g. measure the rate of photosynthesis at various temperatures.

Simone predicted her hair would be worse on date night, based on the theory of sod's law.

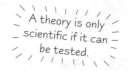

A theory is only scientific if it can be tested.

...Then They **Tell** Everyone About Their **Results**...

The results are **published** — scientists need to let others know about their work. Scientists publish their results in **scientific journals**. These are just like normal magazines, only they contain **scientific reports** (called papers) instead of the latest celebrity gossip.

1) Scientific reports are similar to the **lab write-ups** you do in school. And just as a lab write-up is **reviewed** (marked) by your teacher, reports in scientific journals undergo **peer review** before they're published.

2) The report is sent out to **peers** — other scientists who are experts in the **same area**. They examine the data and results, and if they think that the conclusion is reasonable it's **published**. This makes sure that work published in scientific journals is of a **good standard**.

3) But peer review **can't guarantee** the science is **correct** — other scientists still need to **reproduce** it.

4) Sometimes **mistakes** are made and flawed work is published. Peer review **isn't perfect** but it's probably the best way for scientists to self-regulate their work and to publish **quality reports**.

...Then **Other Scientists** Will **Test** the Theory Too

Other scientists read the published theories and results, and try to **test the theory** themselves. This involves:

• Repeating the **exact same experiments**.

• Using the theory to make **new predictions** and then testing them with **new experiments**.

If the **Evidence** Supports a Theory, It's **Accepted** — for Now

1) If all the experiments in all the world provide good evidence to back it up, the theory is thought of as **scientific 'fact'** (for now).

2) But it will never become **totally indisputable** fact. Scientific **breakthroughs or advances** could provide new ways to question and test the theory, which could lead to **new evidence** that **conflicts** with the current evidence. Then the testing starts all over again...

And this, my friend, is the **tentative nature of scientific knowledge** — it's always **changing** and **evolving**.

The Scientific Process

So scientists need evidence to back up their theories. They get it by carrying out experiments, and when that's not possible they carry out studies. But why bother with science at all? We want to know as much as possible so we can use it to try and improve our lives (and because we're nosy).

Evidence Comes from Lab Experiments...

1) Results from **controlled experiments** in **laboratories** are **great**.
2) A lab is the easiest place to **control variables** so that they're all **kept constant** (except for the one you're investigating).
3) This means you can draw meaningful **conclusions**.

For example, if you're investigating how light intensity affects the rate of photosynthesis you need to keep everything but the light intensity constant, e.g. the temperature, the concentration of carbon dioxide etc.

...and Well-Designed Studies

1) There are things you **can't** investigate in a lab, e.g. whether using a pesticide on farmland affects the number of non-pest species. You have to do a study instead.
2) You still need to try and make the study as controlled as possible to make it **more reliable**. But in reality it's **very hard** to control **all the variables** that **might** be having an effect.
3) You can do things to help, like having a **control** — e.g. an area of similar farmland nearby where the pesticide isn't applied. But you can't easily rule out every possibility.

Having a control reduced the effect of exercise on the study.

See pages 96-98 for more on study design.

Society Makes Decisions Based on Scientific Evidence

1) Lots of scientific work eventually leads to **important discoveries** or breakthroughs that could **benefit humankind**.
2) These results are **used by society** (that's you, me and everyone else) to **make decisions** — about the way we live, what we eat, what we drive, etc.
3) All sections of society use scientific evidence to make decisions, e.g. politicians use it to devise policies and individuals use science to make decisions about their own lives.

Other factors can **influence** decisions about science or the way science is used:

Economic factors

- Society has to consider the **cost** of implementing changes based on scientific conclusions — e.g. the **NHS** can't afford the most expensive drugs without **sacrificing** something else.
- Scientific research is **expensive** so companies won't always develop new ideas — e.g. developing new drugs is costly, so pharmaceutical companies often only invest in drugs that are likely to make them **money**.

Social factors

- **Decisions** affect **people's lives** — E.g. scientists may suggest **banning smoking** and **alcohol** to prevent health problems, but shouldn't **we** be able to **choose** whether **we** want to smoke and drink or not?

Environmental factors

- Scientists believe **unexplored regions** like remote parts of rainforests might contain **untapped drug** resources. But some people think we shouldn't **exploit** these regions because any interesting finds may lead to **deforestation** and **reduced biodiversity** in these areas.

So there you have it — how science works...

Hopefully these pages have given you a nice intro to how science works, e.g. what scientists do to provide you with 'facts'. You need to understand this, as you're expected to know how science works — for the exam and for life.

Photosynthesis and Energy Supply

OK, this isn't the easiest topic to start a book on, but 'cos I'm feeling nice today we'll take it slowly, one bit at a time...

Biological Processes Need Energy

Plant and animal cells **need energy** for biological processes to occur:

- **Plants** need energy for things like **photosynthesis**, **active transport** (e.g. to take in minerals via their roots), **DNA replication**, **cell division** and **protein synthesis**.
- **Animals** need energy for things like **muscle contraction**, maintenance of **body temperature**, **active transport**, **DNA replication**, **cell division** and **protein synthesis**.

Without energy, these biological processes would stop and the plant or animal would die.

Photosynthesis Stores Energy in Glucose

1) **Photosynthesis** is the process where **energy** from **light** is used to **break apart** the **strong bonds** in H_2O molecules — **hydrogen** is **combined** with CO_2 to form **glucose**, and O_2 is **released** into the atmosphere.

2) Photosynthesis occurs in a **series** of **reactions**, but the overall equation is:

$$6CO_2 + 6H_2O + Energy \longrightarrow C_6H_{12}O_6 \text{ (glucose)} + 6O_2$$

3) So, energy is **stored** in the **glucose** until the plants **release** it by **respiration**.

4) Animals obtain glucose by **eating plants** (or **other animals**), then respire the glucose to release energy.

Cells Release Energy from Glucose by Respiration

1) **Plant** and **animal** cells **release energy** from glucose — this process is called **respiration**.

2) This energy is used to power all the **biological processes** in a cell.

3) There are two types of respiration:
   - **Aerobic respiration** — respiration **using oxygen**.
   - **Anaerobic respiration** — respiration **without oxygen**.

4) Aerobic respiration produces **carbon dioxide** and **water**, and releases **energy**. The overall equation is:

$$C_6H_{12}O_6 \text{ (glucose)} + 6O_2 \longrightarrow 6CO_2 + 6H_2O + Energy$$

There's lots more about respiration on pages 54-59.

ATP is the Immediate Source of Energy in a Cell

1) A cell **can't** get its energy **directly** from glucose.

2) So, in respiration, the **energy released** from glucose is used to **make ATP** (adenosine triphosphate). ATP is made from the nucleotide base **adenine**, combined with a **ribose sugar** and **three phosphate groups**.

3) It **carries energy** around the cell to where it's **needed**.

4) **ATP** is **synthesised** from **ADP** and **inorganic phosphate** (P_i) using energy from an **energy-releasing** reaction, e.g. the **breakdown** of **glucose** in **respiration**. The energy is stored as **chemical energy** in the **phosphate bond**. The enzyme **ATP synthase** catalyses this reaction.

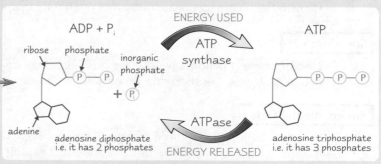

5) ATP **diffuses** to the part of the cell that **needs** energy.

6) Here, it's **broken down** back into **ADP** and **inorganic phosphate** (P_i). Chemical **energy** is **released** from the phosphate bond and used by the cell. **ATPase** catalyses this reaction.

7) The ADP and inorganic phosphate are **recycled** and the process starts again.

Inorganic phosphate (P_i) is just the fancy name for a single phosphate.

Photosynthesis and Energy Supply

You Need to **Know Some Technical Terms** Before You Start

- **Phosphorylation** — **adding phosphate** to a molecule, e.g. **ADP** is phosphorylated to **ATP** (see previous page).
- **Photophosphorylation** — **adding phosphate** to a molecule using **light**.
- **Photolysis** — the **splitting** (lysis) of a molecule using **light** (photo) energy.
- **Hydrolysis** — the **splitting** (lysis) of a molecule using **water** (hydro), e.g. **ATP** is hydrolysed to **ADP**.
- **Redox reactions** — reactions that involve **oxidation** and **reduction**.

1) If something is **reduced** it has **gained electrons** (e^-), and may have **gained hydrogen** or lost oxygen.
2) If something is **oxidised** it has **lost electrons**, and may have **lost hydrogen** or gained oxygen.
3) Oxidation of one molecule **always** involves reduction of another molecule.

One way to remember electron and hydrogen movement is OILRIG. Oxidation Is Loss, Reduction Is Gain.

Photosynthesis Involves Coenzymes

1) A **coenzyme** is a molecule that **aids** the **function** of an **enzyme**.
2) They work by **transferring** a **chemical group** from one molecule to another.
3) A coenzyme used in **photosynthesis** is **NADP**. NADP transfers **hydrogen** from one molecule to another — this means it can **reduce** (give hydrogen to) or **oxidise** (take hydrogen from) a molecule.

Photosynthesis Takes Place in the Chloroplasts of Plant Cells

1) **Chloroplasts are small, flattened organelles** found in **plant cells**.
2) They have a **double membrane** called the **chloroplast envelope**.
3) **Thylakoids** (fluid-filled sacs) are **stacked up** in the chloroplast into structures called **grana** (singular = granum). The grana are **linked** together by bits of thylakoid membrane called **lamellae** (singular = lamella).
4) Chloroplasts contain **photosynthetic pigments** (e.g. **chlorophyll a, chlorophyll b** and **carotene**). These are **coloured substances** that **absorb** the **light energy** needed for photosynthesis. The pigments are found in the **thylakoid membranes** — they're attached to **proteins**. The protein and pigment is called a **photosystem**.
5) There are **two** photosystems used by plants to capture light energy. **Photosystem I** (or PSI) absorbs light best at a wavelength of **700 nm** and **photosystem II** (PSII) absorbs light best at **680 nm**.
6) Contained within the inner membrane of the chloroplast and **surrounding** the thylakoids is a gel-like substance called the **stroma**. It contains **enzymes, sugars, organic acids** and **oil droplets** (which store **non-carbohydrate organic material**).

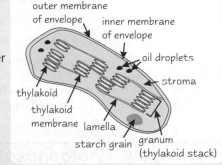

outer membrane of envelope — inner membrane of envelope — oil droplets — stroma — thylakoid — thylakoid membrane — lamella — starch grain — granum (thylakoid stack)

Practice Questions

Q1 Write down three biological processes in animals that need energy.
Q2 What is photosynthesis?
Q3 Give the name of a coenzyme involved in photosynthesis.
Q4 Name two photosynthetic pigments in the chloroplasts of plants.

Exam Question

Q1 ATP is the immediate source of energy inside a cell.
Describe how the synthesis and breakdown of ATP meets the energy needs of a cell. [6 marks]

Oh dear, I've used up all my ATP on these two pages...

Well, I won't beat about the bush, this stuff is pretty tricky... nearly as hard as a cross between Mr T, Hulk Hogan and Arnie. But, with a little patience and perseverance (and plenty of [chocolate] [coffee] [marshmallows] — delete as you wish), you'll get there. Once you've got these pages straight in your head, the next ones will be easier to understand.

The Light-Dependent Reaction

Right, pen at the ready. Check. Brain switched on. Check. Cuppa piping hot. Check. Sweets on standby. Check. Okay, I think you're all sorted to start photosynthesis. Finally, take a deep breath and here we go...

Photosynthesis can be Split into Two Stages

There are actually **two stages** that make up **photosynthesis**:

See p. 8 for loads more information on the Calvin cycle.

1 The Light-Dependent Reaction

1) As the name suggests, this reaction **needs light energy**.

2) It takes place in the **thylakoid membranes** of the chloroplasts.

3) Here, light energy is absorbed by **photosynthetic pigments** in the **photosystems** and converted to **chemical energy**.

4) The light energy is used to add a phosphate group to ADP to form **ATP**, and to reduce NADP to form **reduced NADP**. **ATP transfers energy** and reduced **NADP transfers hydrogen** to the light-independent reaction.

5) During the process H_2O is **oxidised** to O_2.

2 The Light-Independent Reaction

1) This is also called the **Calvin cycle** and as the name suggests it **doesn't use light energy** directly. (But it does **rely** on the **products** of the light-dependent reaction.)

2) It takes place in the **stroma** of the chloroplasts.

3) Here, the **ATP** and **reduced NADP** from the light-dependent reaction supply the **energy** and **hydrogen** to make **glucose** from CO_2.

In the Light-Dependent Reaction ATP is Made by Photophosphorylation

In the light-dependent reaction, the **light energy** absorbed by the photosystems is used for **three** things:

1) Making **ATP** from **ADP** and **inorganic phosphate**. This reaction is called **photophosphorylation** (see p. 5).

2) Making **reduced NADP** from **NADP**.

3) Splitting **water** into **protons** (H^+ ions), **electrons** and **oxygen**. This is called **photolysis** (see p. 5).

The light-dependent reaction actually includes **two types** of **photophosphorylation** — **non-cyclic** and **cyclic**. Each of these processes has **different products**.

Non-cyclic Photophosphorylation Produces ATP, Reduced NADP and O₂

To understand the process you need to know that the photosystems (in the thylakoid membranes) are **linked** by **electron carriers**. Electron carriers are **proteins** that **transfer electrons**. The photosystems and electron carriers form an **electron transport chain** — a **chain** of **proteins** through which **excited electrons flow**. All the processes in the diagrams are happening together — I've just split them up to make it easier to understand.

1 Light energy excites electrons in chlorophyll

- **Light energy** is absorbed by **PSII**.
- The light energy **excites electrons** in **chlorophyll**.
- The electrons move to a **higher energy level** (i.e. they have more energy).
- These high-energy electrons **move along the electron transport chain** to **PSI**.

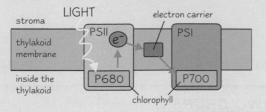

2 Photolysis of water produces protons (H^+ ions), electrons and O₂

- As the excited electrons **from chlorophyll leave PSII** to **move along** the electron transport chain, they must be **replaced**.
- **Light** energy splits **water** into **protons** (H^+ ions), **electrons** and **oxygen**. (So the O_2 in photosynthesis comes from water.)
- The reaction is: $H_2O \longrightarrow 2H^+ + \frac{1}{2}O_2$

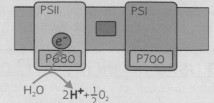

Not all of the electron carriers are shown in these diagrams.

The Light-Dependent Reaction

3) Energy from the excited electrons makes ATP...

- The excited electrons **lose energy** as they **move along** the **electron transport chain**.
- This energy is used to **transport protons into** the **thylakoid** so that the thylakoid has a **higher concentration** of protons than the stroma. This forms a **proton gradient** across the membrane.
- Protons move **down** their concentration gradient, into the stroma, **via** an enzyme called **ATP synthase**. The energy from this movement combines **ADP** and **inorganic phosphate** (P$_i$) to form **ATP**.

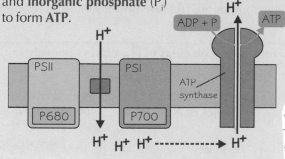

4) ...and generates reduced NADP.

- Light energy is **absorbed** by PSI, which excites the electrons again to an **even higher** energy level.
- Finally, the electrons are **transferred** to **NADP**, along with a **proton** (H$^+$ ion) from the **stroma**, to form **reduced NADP**.

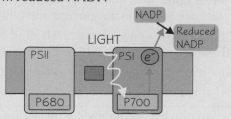

Remember a 'proton' is just another word for a hydrogen ion (H$^+$).

Chemiosmosis is the name of the process where the movement of H$^+$ ions across a membrane generates ATP. This process also occurs in respiration (see p. 57).

Cyclic Photophosphorylation *Only Produces* ATP

Cyclic photophosphorylation **only uses PSI**. It's called 'cyclic' because the electrons from the chlorophyll molecule **aren't** passed onto NADP, but are **passed back** to PSI via electron carriers. This means the electrons are **recycled** and can repeatedly flow through PSI. This process doesn't produce any reduced NADP or O$_2$ — it **only produces** small amounts of **ATP**.

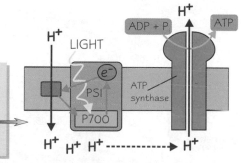

Practice Questions

Q1 Name the two stages of photosynthesis.

Q2 What three substances does non-cyclic photophosphorylation produce?

Q3 Which photosystem is involved in cyclic photophosphorylation?

Exam Question

Q1 The diagram on the right shows the light-dependent reaction of photosynthesis.
  a) Where precisely in a plant does the light-dependent reaction of photosynthesis occur? [1 mark]
  b) What is A? [1 mark]
  c) Describe process B and explain its purpose. [4 marks]
  d) What is reactant D? [1 mark]

Cramming in the early hours before an exam is definitely light-dependent...

I know I said the first two pages were tricky, but these two are also pretty mental — once you get into it they're less scary than they look though (scout's promise). But don't panic if you don't get it the first time you read it — try reading it again. I'd have a go at drawing out the diagrams too if I were you, so you know exactly what's going on in photophosphorylation.

The Light-Independent Reaction

Don't worry, you're over the worst of photosynthesis now. Instead of electrons flying around, there's a nice cycle of reactions to learn. What more could you want from life? Money, fast cars and nice clothes have nothing on this...

The **Light-Independent** Reaction is also called the **Calvin Cycle**

1) The Calvin cycle takes place in the **stroma** of the chloroplasts.

2) It involves the **reduction of CO_2** — CO_2 is **combined** with **hydrogen** during a **series** of **reactions**. This makes a molecule called **triose phosphate**. Triose phosphate can then used to make **glucose** and other **useful organic substances** (see below).

3) There are a few steps in the cycle, and it needs **energy** and **H^+ ions** to keep it going. These are provided by **ATP** and **reduced NADP** from the **light-dependent reaction**.

4) The reactions are linked in a **cycle**, which means the starting compound, **ribulose bisphosphate**, is **regenerated**.

Here's what happens at each stage in the cycle:

> *The Calvin cycle is also called carbon fixation, because carbon from CO_2 is 'fixed' into an organic molecule.*

1 **Carbon dioxide is combined with ribulose bisphosphate to form two molecules of glycerate 3-phosphate**

- CO_2 enters the leaf through the **stomata** and diffuses into the **stroma** of the chloroplast.
- Here, it's combined with **ribulose bisphosphate (RuBP)**, a **5-carbon** compound. This gives an **unstable 6-carbon** compound, which quickly breaks down into **two** molecules of a **3-carbon** compound called **glycerate 3-phosphate (GP)**.
- **Ribulose bisphosphate carboxylase (rubisco)** catalyses the reaction between CO_2 and **ribulose bisphosphate**.

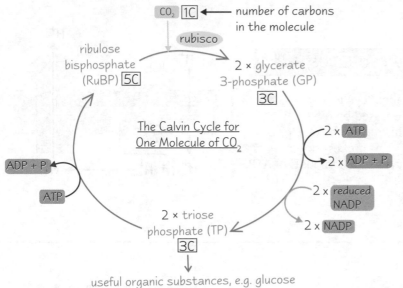

number of carbons in the molecule

The Calvin Cycle for One Molecule of CO_2

2 **ATP and reduced NADP are required for the reduction of GP to triose phosphate**

- Now **ATP**, from the **light-dependent** reaction, **provides energy** to **reduce** the **3-carbon** compound, **GP**, to a **different** 3-carbon compound called **triose phosphate (TP)**.
- This **reduction** reaction also requires **H^+ ions**, which come from **reduced NADP** (also from the **light-dependent reaction**). Reduced NADP is **recycled** to **NADP**.
- **Triose phosphate** is then converted into many **useful organic compounds**, e.g. glucose (see below).

> *Triose phosphate is also called glyceraldehyde 3-phosphate (GALP).*

3 **Ribulose bisphosphate is regenerated**

- **Five** out of every **six** molecules of **TP** produced in the cycle aren't used to make hexose sugars, but to **regenerate RuBP**.
- Regenerating RuBP uses the **rest** of the **ATP** produced by the **light-dependent reaction**.

TP and **GP** are **Converted** into **Useful Organic Substances** like **Glucose**

The Calvin cycle is the starting point for making **all** the organic substances a plant needs. **Triose phosphate** (TP) and **glycerate 3-phosphate** (GP) molecules are used to make **carbohydrates**, **lipids**, **proteins** and **nucleic acids**:

- **Carbohydrates** — **simple sugars** (e.g. glucose) are made by joining **two triose phosphate molecules** together, and **polysaccharides** (e.g. starch and cellulose) are made by joining **hexose sugars** together in **different ways**.
- **Lipids** — these are made using **glycerol**, which is synthesised from **triose phosphate**, and **fatty acids**, which are synthesised from **glycerate 3-phosphate**.
- **Amino acids** — some amino acids are made from **glycerate 3-phosphate**.
- **Nucleic acids** — the sugar in **RNA** (**ribose**) is made using **triose phosphate**.

The Light-Independent Reaction

The Calvin Cycle Needs to Turn Six Times to Make One Glucose Molecule

Here's the reason why:

1) **Three turns** of the cycle produces **six** molecules of **triose phosphate** (TP), because two molecules of TP are made for every one CO_2 molecule used.

2) **Five** out of **six** of these TP molecules are used to **regenerate ribulose bisphosphate** (RuBP).

3) This means that for **three turns** of the cycle only **one TP** is produced that's used to make glucose.

4) Glucose has **six carbons** though, so **two TP** molecules are needed to form one glucose molecule.

5) This means the cycle must turn **six times** to produce **two molecules** of **TP** that can be used to make **one glucose molecule**.

6) Six turns of the cycle need **18 ATP** and **12 reduced NADP** from the light-dependent reaction.

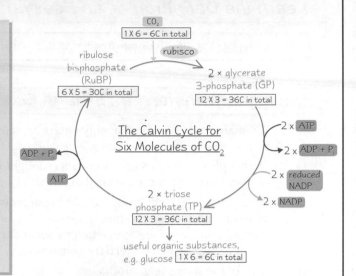

This might seem a bit inefficient, but it keeps the cycle going and makes sure there's always **enough RuBP** ready to combine with CO_2 taken in from the atmosphere.

Morag had to turn one million times to make a sock... two million for a scarf.

The Structure of a Chloroplast is Adapted for Photosynthesis

1) The **chloroplast envelope** keeps the **reactants** for photosynthesis **close** to their **reaction sites**.

2) The **thylakoids** have a **large surface area** to allow as much **light energy** to be **absorbed** as possible.

3) **Lots** of **ATP synthase** molecules are present in the thylakoid membranes to **produce ATP** in the light-dependent reaction.

4) The **stroma** contains all the **enzymes**, **sugars** and **organic acids** for the light-independent reaction to take place.

Practice Questions

Q1 Where in the chloroplasts does the light-independent reaction occur?
Q2 Name two organic substances made from triose phosphate.
Q3 How many CO_2 molecules need to enter the Calvin cycle to make one glucose molecule?
Q4 Describe two ways in which a chloroplast is adapted for photosynthesis.

Exam Question

Q1 Rubisco is an enzyme that catalyses the first reaction of the Calvin cycle.
CA1P is an inhibitor of rubisco.
a) Describe how triose phosphate is produced in the Calvin cycle. [6 marks]
b) Briefly explain how ribulose bisphosphate (RuBP) is regenerated in the Calvin cycle. [2 marks]
c) Explain the effect that CA1P would have on glucose production. [3 marks]

Calvin cycles — bikes made by people that normally make pants...

Next thing we know there'll be male models swanning about in their pants riding highly fashionable bikes. Sounds awful I know, but let's face it, anything would look better than cycling shorts. Anyway, it would be a good idea to go over these pages a couple of times — you might not feel as if you can fit any more information in your head, but you can, I promise.

Energy Transfer and Productivity

Ecology — whether you love it or hate it, it's time to get in touch with nature...

Learn the Definition of an Ecosystem

An **ecosystem** is all the **organisms** living in a particular area and all the **non-living** (abiotic) factors.

Energy is Transferred Through Ecosystems

1) The **main route** by which energy **enters** an ecosystem is **photosynthesis** (e.g. by plants, see p. 4). (Some energy enters sea ecosystems when bacteria use chemicals from deep sea vents as an energy source.)

2) During photosynthesis plants **convert sunlight energy** into a form that can be **used** by other organisms — plants are called **producers** (because they produce **organic molecules** using sunlight energy).

3) Energy is **transferred** through the **living organisms** of an ecosystem when organisms **eat** other organisms, e.g. producers are eaten by organisms called **primary consumers**. Primary consumers are then eaten by **secondary consumers** and secondary consumers are eaten by **tertiary consumers**.

4) Each of the stages (e.g. producers, primary consumers) are called **trophic levels**.

5) **Food chains** and **food webs** show how energy is **transferred** through an ecosystem.

6) **Food chains** show **simple lines** of energy transfer.

7) **Food webs** show **lots** the **food chains** in an ecosystem and how they **overlap**.

8) Energy locked up in the things that **can't be eaten** (e.g. bones, faeces) gets recycled back into the ecosystem by microorganisms called **decomposers** — they **break down dead** or **undigested** material.

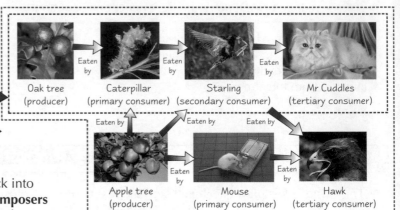

Oak tree (producer) → Eaten by → Caterpillar (primary consumer) → Eaten by → Starling (secondary consumer) → Eaten by → Mr Cuddles (tertiary consumer)

Apple tree (producer) → Eaten by → Mouse (primary consumer) → Eaten by → Hawk (tertiary consumer)

Not All Energy gets Transferred to the Next Trophic Level

1) **Not all** the energy (e.g. from sunlight or food) that's available to the organisms in a trophic level is **transferred** to the **next** trophic level — around **90%** of the **total available energy** is **lost** in various ways.

2) Some of the available energy (**60%**) is **never taken in** by the organisms in the first place. For example:

   • Plants **can't use** all the light energy that reaches their **leaves**, e.g. some is the **wrong wavelength**, some is **reflected**, and some **passes straight through** the leaves.

   • Some sunlight can't be used because it hits parts of the plant that **can't photosynthesise**, e.g. the bark of a tree.

   • Some **parts** of food, e.g. **roots** or **bones**, **aren't eaten** by organisms so the energy isn't taken in.

   • Some parts of food are **indigestible** so **pass through** organisms and come out as **waste**, e.g. **faeces**.

3) The rest of the available energy (**40%**) is **taken in** (absorbed) — this is called the **gross productivity**. But not all of this is available to the next trophic level either.

   • **30%** of the **total energy** available (75% of the gross productivity) is **lost to the environment** when organisms use energy produced from **respiration** for **movement** or body **heat**. This is called **respiratory loss**.

   • **10%** of the **total energy** available (25% of the gross productivity) becomes **biomass** (e.g. it's **stored** or used for **growth**) — this is called the **net productivity**.

4) **Net productivity** is the amount of energy that's **available** to the **next trophic level**.

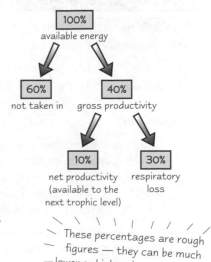

100% available energy

60% not taken in   40% gross productivity

10% net productivity (available to the next trophic level)   30% respiratory loss

These percentages are rough figures — they can be much lower or higher depending on the food chain and trophic level.

Energy Transfer and Productivity

Productivity and Energy Transfer Between Trophic Levels can be Calculated

You can **calculate** the **net productivity** of a trophic level when you know the **gross productivity** and the **respiratory losses** of the trophic level. Here's the **equation**:

Here's an example of how **net productivity** is **calculated**:

$$\text{net productivity} = \text{gross productivity} - \text{respiratory loss}$$

The rabbits in an ecosystem receive **20 000 kJm^{-2}yr^{-1}** of energy, but don't take in **12 000 kJm^{-2}yr^{-1}** of it, so their gross productivity is **8000 kJm^{-2}yr^{-1}** (20 000 – 12 000). They lose **6000 kJm^{-2}yr^{-1}** using energy from **respiration**. You can use this to **calculate** the **net productivity** of the rabbits:

$$\text{net productivity} = 8000 - 6000$$
$$= 2000 \text{ kJm}^{-2}\text{yr}^{-1}$$

The outlook was grim for Jimmy and his crew — net productivity was at an all-time low.

You might be asked to **calculate** how **efficient energy transfer** from one trophic level to another is. Here's the **equation**:

$$\text{\% efficiency of energy transfer between trophic levels} = \frac{\text{net productivity of a level}}{\text{net productivity of previous level}} \times 100$$

Here's an example of how the **efficiency of energy transfer** between trophic levels is **calculated**:

The rabbits receive **20 000 kJm^{-2}yr^{-1}**, and their **net productivity** is **2000 kJm^{-2}yr^{-1}**. So the **percentage efficiency of energy transfer** is:

$$2000 \div 20\,000 = 0.1$$
$$0.1 \times 100 = 10\%$$

Primary Productivity can be Calculated too

When you're talking about **producers**, net productivity is called **net primary productivity** (NPP) and gross productivity is called **gross primary productivity** (GPP), so the equation is:

$$\text{NPP} = \text{GPP} - \text{plant respiration}$$

NPP is lower when it's cold or there's not a lot of light, as photosynthesis is slower.

Here's an example of how **net primary productivity** is **calculated**:

The **grass** in an ecosystem receives **950 000 kJm^{-2}yr^{-1}** of sunlight energy. It doesn't take in **931 000 kJm^{-2}yr^{-1}** of the energy received, so the gross primary productivity of the grass is **19 000 kJm^{-2}yr^{-1}** (950 000 – 931 000). The grass loses **8000 kJm^{-2}yr^{-1}** using energy from **respiration**. You can use this to **calculate** the **net primary productivity** of the grass:

$$\text{net primary productivity} = 19\,000 - 8000$$
$$= 11\,000 \text{ kJm}^{-2}\text{yr}^{-1}$$

Practice Questions

Q1 What is the main route by which energy enters ecosystems?

Q2 Explain the term gross productivity.

Q3 What is the equation for net primary productivity?

Grass 13 883 kJm^{-2}yr^{-1}	→	Arctic hare 2345 kJm^{-2}yr^{-1}	→	Arctic fox 137 kJm^{-2}yr^{-1}

Exam Question

Q1 The diagram above shows the net productivity of different trophic levels in a food chain.

a) Explain why the net productivity of the Arctic hare is less than the net productivity of the grass. [4 marks]

b) Calculate the percentage efficiency of energy transfer from the Arctic hare to the Arctic fox. [2 marks]

Boy, do I need an energy transfer this morning...

It's really important to remember that energy transfer through an ecosystem isn't 100% efficient — most gets lost along the way so the next organisms don't get all the energy. Make sure you can calculate productivity and the efficiency of energy transfers — you might not like maths, but they're easy marks in the exam. Then prepare your brain for some more ecology...

Factors Affecting Abundance and Distribution

Now you need to get to grips with what affects where organisms hang out and how many there are in an area.
(Warning: contains upsetting information about cute bunny-wunnys being eaten.)

You Need to **Learn Some Definitions** to get you **Started**

Habitat	—	The **place** where an organism **lives**, e.g. a rocky shore or a field.
Population	—	**All** the organisms of **one species** in a **habitat**.
Population size	—	The **number of individuals** of **one species** in a **particular area**.
Community	—	Populations of **different species** in a habitat make up a **community**.
Abiotic factors	—	The **non-living** features of the ecosystem, e.g. **temperature** and **availability of water**.
Biotic factors	—	The **living** features of the ecosystem, e.g. the presence of **predators** or **food**.
Abundance	—	The **number of individuals** of **one species** in a **particular area**. (It's the **same** as **population size**.)
Distribution	—	**Where** a species is within a **particular area**.

Being a member of the undead made it hard for Mumra to know whether he was a living or a non-living feature of the ecosystem.

Population Size (Abundance) Varies Because of Abiotic Factors...

1) The **population size** of any species **varies** because of **abiotic** factors, e.g. the amount of **light**, **water** or **space** available, the **temperature** of their surroundings or the **chemical composition** of their surroundings.

2) When abiotic conditions are **ideal** for a species, organisms can **grow fast** and **reproduce successfully**.

> For example, when the temperature of a mammal's surroundings is the ideal temperature for **metabolic reactions** to take place, they don't have to **use up** as much energy **maintaining** their **body temperature**. This means more energy can be used for **growth** and **reproduction**, so their population size will **increase**.

3) When abiotic conditions **aren't ideal** for a species, organisms **can't** grow as **fast** or reproduce as **successfully**.

> For example, when the temperature of a mammal's surroundings is significantly **lower** or **higher** than their **optimum** body temperature, they have to **use** a lot of **energy** to maintain the right **body temperature**. This means less energy will be available for **growth** and **reproduction**, so their population size will **decrease**.

Abiotic and biotic factors are sometimes called abiotic and biotic conditions.

...and Because of Biotic Factors

(1) Interspecific Competition — Competition Between Different Species

1) Interspecific competition is when organisms of **different species compete** with each other for the **same resources**, e.g. **red** and **grey** squirrels compete for the same **food sources** and **habitats** in the **UK**.

2) Interspecific competition between two species can mean that the **resources available** to **both** populations are **reduced**, e.g. if they share the **same** source of food, there will be **less** available to both of them. This means both populations will be **limited** by a lower amount of food. They'll have less **energy** for **growth** and **reproduction**, so the population sizes will be **lower** for both species. E.g. in areas where both **red** and **grey** squirrels live, both populations are **smaller** than they would be if there was **only one** species there.

Factors Affecting Abundance and Distribution

2) Intraspecific Competition — Competition Within a Species

Intraspecific competition is when organisms of the **same species compete** with each other for the **same resources**.

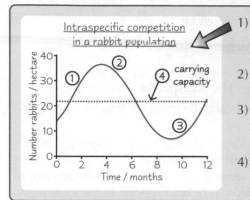

Intraspecific competition in a rabbit population

1) The **population** of a species (e.g. rabbits) **increases** when resources are **plentiful**. As the population increases, there'll be **more** organisms competing for the **same amount** of **space** and **food**.

2) Eventually, resources such as food and space become **limiting** — there **isn't enough** for all the organisms. The population then begins to **decline**.

3) A **smaller** population then means that there's **less competition** for space and food, which is **better** for **growth** and **reproduction** — so the population starts to **grow** again.

4) The **maximum stable population size** of a species that an ecosystem can **support** is called the **carrying capacity**.

3) Predation — Predator and Prey Population Sizes are Linked

Predation is where an organism (the predator) kills and eats another organism (the prey), e.g. lions kill and eat (**predate** on) buffalo. The **population sizes** of predators and prey are **interlinked** — as the population of one **changes**, it **causes** the other population to **change**:

1) As the **prey** population **increases**, there's **more food** for predators, so the **predator** population **grows**. E.g. in the graph on the right the **lynx** population **grows** after the **snowshoe hare** population has **increased** because there's **more food** available.

2) As the **predator** population **increases**, **more prey** is **eaten** so the **prey** population then begins to **fall**. E.g. **greater numbers** of lynx eat lots of snowshoe hares, so their population **falls**.

3) This means there's **less food** for the **predators**, so their population **decreases**, and so on. E.g. **reduced** snowshoe hare numbers means there's **less food** for the lynx, so their population **falls**.

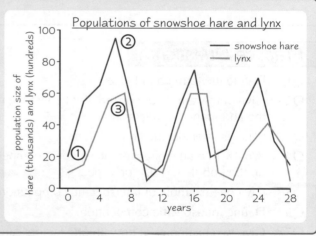

Populations of snowshoe hare and lynx

Predator-prey relationships are usually more **complicated** than this though because there are **other factors** involved, like availability of **food** for the **prey**. E.g. it's thought that the population of snowshoe hare initially begins to **decline** because there's **too many** of them for the amount of **food available**. This is then **accelerated** by **predation** from the lynx.

Distribution Varies Because of Abiotic Factors...

Organisms can **only exist** where the **abiotic** factors they can **survive in** exist. For example:

- Some **plants** only grow on **south-facing slopes** in the northern hemisphere because that's where **solar input** (light intensity) is **greatest**.
- Some plants **don't** grow near to the **shoreline** because the **soil** is **too saline** (salty).
- **Large trees can't** grow in **polar regions** because the **temperature** is **too low**.

...and Because of Biotic Factors

Interspecific competition can affect the **distribution** of species. If **two** species are competing but one is **better adapted** to its surroundings than the other, the less well adapted species is likely to be **out-competed** — it **won't** be able to **exist** alongside the better adapted species. E.g. since the introduction of the **grey squirrel** to the UK, the native **red squirrel** has **disappeared** from large areas. The grey squirrel has a better chance of **survival** because it's **larger** and can store **more fat** over winter.

Factors Affecting Abundance and Distribution

Every Species Occupies a *Different Niche*

Don't get confused between habitat (where a species lives) and niche (what it does in its habitat).

1) A **niche** is the **role** of a species within its habitat. It includes:

  - Its **biotic** interactions — e.g. the organisms it **eats**, and those it's **eaten by**.
  - Its **abiotic** interactions — e.g. the **oxygen** an organism breathes in, and the **carbon dioxide** it breathes out.

2) Every species has its own **unique niche** — a niche can only be occupied by **one species**.

3) It may **look** like **two species** are filling the **same niche** (e.g. they're both eaten by the same species), but there'll be **slight differences** (e.g. variations in what they eat).

4) The **abundance** of different species can be **explained** by the niche concept — two species occupying **similar** niches will **compete** (e.g. for a **food source**), so **fewer individuals** of **both** species will be able to survive in the area. For example, common and soprano pipistrelle bats feed on the **same insects**. This means the **amount of food** available to both species is **reduced**, so there will be **fewer individuals** of **both** species in the same area.

5) The **distribution** of different species can also be **explained** by the niche concept — organisms can only **exist** in habitats where all the **conditions** that make up their **role exist**. For example, the soprano pipistrelle bat feeds on **insects** and lives in **farmland**, **open woodland**, **hedgerows** and **urban areas** — it **couldn't exist** in a **desert** because there are **different insects** and **no woodland**.

Practice Questions

Q1 Define the term abundance.

Q2 Give one example of how an abiotic factor can affect the abundance of organisms.

Q3 What is interspecific competition?

Q4 What will be the effect of interspecific competition on the population size of a species?

Q5 Give one example of interspecific competition.

Q6 Define intraspecific competition.

Q7 Give two examples of how abiotic factors can affect the distribution of organisms.

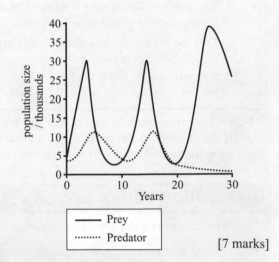

Exam Questions

Q1 The graph on the right shows the population size of a predator species and a prey species over a period of 30 years.

  a) Using the graph, describe and explain how the population sizes of the predator and prey species vary over the first 20 years. [7 marks]

  b) The numbers of species B declined after year 20 because of a disease. Describe and explain what happened to the population of species A. [4 marks]

Q2 Two species of lizard (X and Y) live in the same area. Both feed on the same insects and are eaten by the same predator species. Species X feeds mainly during the morning and species Y feeds mainly during the afternoon.

  a) Explain the term 'niche'. [2 marks]

  b) Explain how having a similar niche affects the abundance of each lizard species in the area. [2 marks]

Predator-prey relationships — they don't usually last very long...

You'd think they could have come up with names a little more different than inter- and intraspecific competition. I always remember it as int-er means diff-er-ent species. The factors that affect abundance and distribution are divided up nicely for you here into abiotic and biotic factors — just like predators like to nicely divide up their prey into bitesize chunks.

Investigating Populations and Abiotic Factors

Examiners aren't happy unless you're freezing to death in the rain in a field somewhere in the middle of nowhere. Still, it's better than being stuck in the classroom being bored to death learning about fieldwork techniques...

You need to be able to **Investigate Populations** of **Organisms**

Investigating **populations** of organisms involves looking at the **abundance** and **distribution** of species in a particular **area**.

1) **Abundance** — the **number of individuals** of **one species** in a **particular area** (i.e. the **population size**). The abundance of **mobile organisms** and **plants** can be estimated by simply counting the **number** of individuals in samples taken. **Percentage cover** can also be used to measure the abundance of plants — this is **how much** of the area you're investigating is **covered** by a species.

2) **Distribution** — this is **where** a particular species is within the **area you're investigating**.

You need to take a **Random Sample** from the **Area You're Investigating**

Most of the time it would be too **time-consuming** to measure the **number of individuals** and the **distribution** of every species in the **entire area** you're investigating, so instead you take **samples**:

1) **Choose an area** to **sample** — a **small** area **within** the area being investigated.

2) Samples should be **random** to **avoid bias**, e.g. if you were investigating a field you could pick random sample sites by dividing the field into a **grid** and using a **random number generator** to select **coordinates**.

3) Use an **appropriate technique** to take a sample of the population (see pages 15-16).

4) **Repeat** the process, taking as many samples as possible. This gives a more **reliable** estimate for the **whole area**. (If your **results aren't reliable** your **conclusion won't** be **valid**.)

5) The **number of individuals** for the **whole area** can then be **estimated** by taking an **average** of the data collected in each sample and **multiplying** it by the size of the whole area. The **percentage cover** for the whole area can be estimated by taking the average of all the samples.

Finally! 26 542 981 poppies. What do you mean I didn't need to count them all?

Frame Quadrats can be used to **Investigate Plant Populations**

1) A **frame quadrat** is a **square** frame divided into a **grid** of 100 **smaller squares** by strings attached across the frame.

2) They're **placed on the ground** at **random points** within the area you're investigating. This can be done by selecting **random coordinates** (see above).

3) The **number of individuals** of each species is recorded in **each quadrat**.

4) The **percentage cover** of a **plant species** can also be measured by counting how much of the quadrat is **covered** by the plant species — you count a square if it's **more than half-covered**. Percentage cover is a **quick** way to investigate populations because you **don't** have to **count** all the **individual** plants.

5) Frame quadrats are useful for **quickly** investigating areas with species that **fit** within a **small quadrat** — most frame quadrats are **1 m by 1 m**.

6) Areas with **larger plants** and **trees** need **very large** quadrats. Large quadrats **aren't** always in a frame — they can be marked out with a **tape measure**.

the area of this quadrat is 0.25 m²
0.5 m
0.5 m

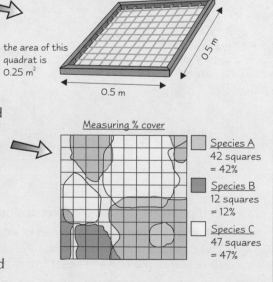

Measuring % cover

Species A
42 squares
= 42%

Species B
12 squares
= 12%

Species C
47 squares
= 47%

Investigating Populations and Abiotic Factors

Point Quadrats *can also be used to* Investigate Plant Populations

1) A **point quadrat** is a **horizontal bar** on **two legs** with a series of holes at set intervals along its length.

2) Point quadrats are **placed on the ground** at **random points** within the area you're investigating.

3) **Pins** are dropped through the holes in the frame and **every plant** that each pin **touches** is **recorded**. If a pin touches several **overlapping** plants, **all** of them are recorded.

4) The **number of individuals** of each species is recorded in **each quadrat.**

5) The **percentage cover** of a species can also be measured by calculating the **number of times** a pin has touched a species as a **percentage** of the **total number** of pins dropped.

6) Point quadrats are especially useful in areas where there's lots of **dense vegetation** close to the ground.

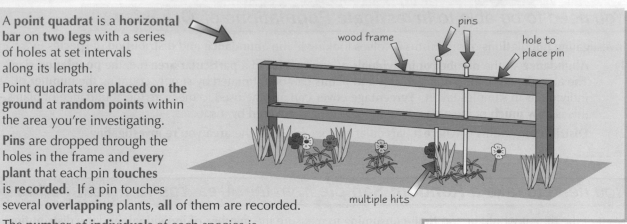

It's a horizontal bar with two legs alright, but where do we put the pins?

Transects *are used to* Investigate *the* Distribution *of* Plant Populations

You can use **lines** called **transects** to help find out how plants are **distributed across** an area, e.g. how species **change** from a hedge towards the middle of a field. There are **three** types of transect:

1) **Line transects** — a **tape measure** is placed **along** the transect and the species that **touch** the tape measure are **recorded.**

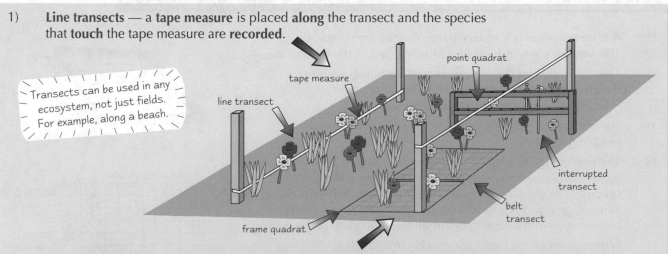

Transects can be used in any ecosystem, not just fields. For example, along a beach.

2) **Belt transects** — data is collected along the transect using **frame quadrats** placed **next to** each other.

3) **Interrupted transects** — instead of investigating the **whole transect** of either a line or a belt, you can take **measurements** at **intervals**. E.g. by placing **point quadrats** at **right angles** to the direction of the transect at **set intervals** along its length, such as **every 2 m.**

Investigating Populations and Abiotic Factors

You can also Measure Different Abiotic Factors

The **abundance** and **distribution** of organisms is **affected** by **abiotic** factors.
You need to know how to **measure** some of them:

1) **Climate** — the **weather** of a region:

   - **Temperature** is measured using a **thermometer**.
   - **Rainfall** is measured using a **rain gauge** — a **funnel** attached to a **measuring cylinder**. The rain **falls into** the funnel and **runs down** into the measuring cylinder. The **volume** of water collected over a **period of time** can be measured.
   - **Humidity** (the amount of **water vapour** in the air) is measured using an electronic **hygrometer**.

2) **Oxygen availability** — this only needs to be measured in **aquatic habitats**. The **amount** of oxygen **dissolved** in the water is measured using an electronic device called an **oxygen sensor**.

3) **Solar input** (light intensity) is measured using an electronic device called a **light sensor**.

4) **Edaphic factors** (**soil** conditions):

   - **pH** is measured using **indicator liquid** — a **sample** of the soil is **mixed** with an indicator liquid that **changes colour** depending on the pH. The colour is matched against a **chart** to determine the pH of the soil. Electronic **pH monitors** can also be used.
   - **Moisture content** — the **mass** of a sample of soil is measured **before** and **after** being **dried out** in an **oven at 80-100 °C** (until it reaches a **constant mass**). The difference in mass as a **percentage** of the **original** mass of the soil is then calculated. This shows the water content of the soil sample.

5) **Topography** — the **shape** and **features** of the Earth's surface:

   - **Relief** (how the **height** of the land changes across a surface) can be measured by taking **height readings** using a **GPS** (global positioning system) device at **different points** across the surface. You can also use **maps** with **contour lines** (lines that join points that are the same height).
   - **Slope angle** (how **steep** a slope is) is measured using a **clinometer**. A simple clinometer is just a piece of **string** with a **weight** on the end attached to the centre of a **protractor**. You **point** the flat edge of the protractor **up the hill**, and read the slope angle off the protractor.
   - **Aspect** (the **direction** a slope is facing) is measured using a **compass**.

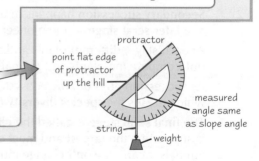

Practice Questions

Q1   Define abundance.

Q2   What does percentage cover show?

Q3   Explain why samples of a population are taken.

Q4   Briefly describe how belt transects are different from line transects.

Exam Question

Q1   A student wants to sample a population of daffodils in a field.

   a)   Describe how she could investigate the percentage cover of daffodils in the field using frame quadrats.   [3 marks]

   b)   The abundance of daffodils can be affected by different abiotic factors.
         Describe two ways to measure the climate in the field.   [2 marks]

What did the quadrat say to the policeman — I've been framed...

If you want to know what it's really like doing these investigations then read these pages outside in the pouring rain. Doing it while you're tucked up in a nice warm, dry exam hall won't seem so bad after that, take my word for it.

Succession

Now that you're a pro on investigating the abundance and distribution of organisms within an ecosystem, you need to know how an ecosystem can change over time.

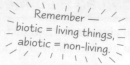

Remember — biotic = living things, abiotic = non-living.

Succession *is the* Process of Ecosystem Change

Succession is the process by which an **ecosystem changes** over **time**. The **biotic conditions** (e.g. **plant** and **animal communities**) change as the **abiotic conditions** change (e.g. **water** availability). There are **two** types of succession:

1) **Primary succession** — this happens on land that's been **newly formed** or **exposed**, e.g. where a **volcano** has erupted to form a **new rock surface**, or where **sea level** has **dropped** exposing a new area of land. There's **no soil** or **organic material** to start with, e.g. just bare rock.

2) **Secondary succession** — this happens on land that's been **cleared** of all the **plants**, but where the **soil remains**, e.g. after a **forest fire** or where a forest has been **cut down by humans**.

Succession *Occurs in* Stages *called* Seral Stages

1) **Primary succession** starts when species **colonise** a new land surface. **Seeds** and **spores** are blown in by the **wind** and begin to **grow**. The **first species** to colonise the area are called **pioneer species** — this is the **first seral stage**.

   - The **abiotic conditions** are **hostile** (harsh), e.g. there's no soil to **retain water**. Only pioneer species **grow** because they're **specialised** to cope with the harsh conditions, e.g. **marram grass** can grow on sand dunes near the sea because it has **deep roots** to get water and can **tolerate** the salty environment.

   - The pioneer species **change** the **abiotic conditions** — they **die** and **microorganisms decompose** the dead **organic material** (**humus**). This forms a **basic soil**.

   - This makes conditions **less hostile**, e.g. the basic soil helps to **retain water**, which means **new organisms** can move in and grow. These then die and are decomposed, adding **more** organic material, making the soil **deeper** and **richer in minerals**. This means **larger plants** like **shrubs** can start to grow in the deeper soil, which retains **even more** water.

2) **Secondary succession** happens in the **same way**, but because there's already a **soil layer** succession starts at a **later seral stage** — the pioneer species in secondary succession are **larger plants**, e.g. shrubs.

3) At each stage, **different** plants and animals that are **better adapted** for the improved conditions move in, **out-compete** the plants and animals that are already there, and become the **dominant species** in the ecosystem.

4) As succession goes on, the ecosystem becomes **more complex**. New species move in **alongside** existing species, which means the **species diversity** (the number of **different species** and the **abundance** of each species) **increases**.

5) The **final seral stage** is called the **climax community** — the ecosystem is supporting the **largest** and **most complex** community of plants and animals it can. It **won't change** much more — it's in a **steady state**.

This example shows primary succession on bare rock, but succession also happens on sand dunes, salt marshes and even on lakes.

Example of primary succession — bare rock to woodland

1) **Pioneer species colonise** the rocks. E.g. **lichens** grow on and **break down** rocks, **releasing minerals**.

2) The lichens **die** and are **decomposed** helping to form a **thin soil**, which thickens as more **organic material** is formed. This means other species such as **mosses** can **grow**.

3) **Larger plants** that need **more water** can move in as the soil **deepens**, e.g. **grasses** and **small flowering plants**. The soil **continues to deepen** as the larger plants die and are decomposed.

4) **Shrubs, ferns** and **small trees** begin to grow, **out-competing** the grasses and smaller plants to become the **dominant** species. **Diversity increases**.

5) Finally, the soil is **deep** and **rich** enough in **nutrients** to support **large trees**. These become the dominant species, and the **climax community** is formed.

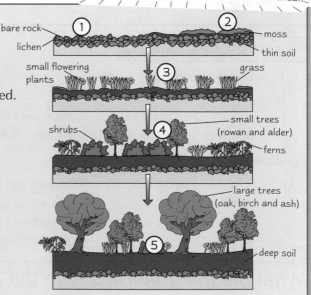

Succession

Different Ecosystems have Different Climax Communities

Which species make up the climax community depends on what the **climate's** like in an ecosystem. The climax community for a **particular** climate is called its **climatic climax**. For example:

> In a **temperate climate** there's **plenty** of **available water**, **mild temperatures** and not much **change** between the seasons. The climatic climax will contain **large trees** because they **can grow** in these conditions once **deep soils** have developed. In a **polar climate** there's **not much available water**, temperatures are **low** and there are **massive changes** between the seasons. Large trees **won't ever** be able to grow in these conditions, so the climatic climax contains only **herbs** or **shrubs**, but it's still the **climax community**.

Succession can be Prevented

Human activities can **prevent succession**, stopping the normal climax community from **developing**. When succession is stopped **artificially** like this, the climax community is called a **plagioclimax**. For example:

> A **regularly mown** grassy field **won't develop** woody plants, even if the climate of the ecosystem could support them. The **growing points** of the woody plants are **cut off** by the lawnmower, so larger plants **can't establish** themselves — only the grasses can **survive** being mowed, so the **climax community** is a **grassy field**.

Man had been given a mighty weapon with which they would tame the forces of nature.

Practice Questions

Q1 What is the difference between primary and secondary succession?

Q2 What is the name given to species that are the first to colonise an area during succession?

Q3 As succession continues what happens to the species diversity in the area?

Q4 Briefly describe an example of primary succession.

Q5 What is meant by a climax community?

Q6 Give an example of how succession can be prevented.

Exam Question

Q1 A farmer has a field where he plants crops every year. When the crops are fully grown he removes them all and then ploughs the field (churns up all the plants and soil so the field is left as bare soil). The farmer has decided not to plant crops or plough the field for several years.

a) Describe, in terms of succession, what will happen in the field over time. [6 marks]

b) Explain why succession doesn't usually take place in the farmer's field. [2 marks]

Revision succession — bare brain to a woodland of knowledge...

When answering questions on succession, examiners are pretty keen on you using the right terminology — that means saying "pioneer species" instead of "the first plants to grow there". This stuff's all quite wordy, but the concept of succession is simple enough — some plants start growing, change the environment so it's less hostile, then others can move in.

Introduction to Global Warming

A2 level student, meet global warming — take a few pages to get to know each other...

Global Warming is the Recent Rise in Global Temperature

1) **Global warming** is the term used for the **rapid increase** in **global temperature** over the **last century**.

2) It's a type of **climate change** — a significant change in the **weather** of a region over a period of at least **several decades**.

3) Global warming also **causes other types** of climate change, e.g. changing **rainfall patterns** and **seasonal cycles**.

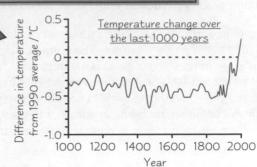

You Need to be able to Interpret Evidence for Global Warming

There are **different types** of **evidence** that can be used to show that global warming **is happening**. You need to be able to interpret three types of evidence for global warming:

① Temperature Records

1) Since the 1850s **temperature** has been **measured** around the world using **thermometers**.

2) This gives a **reliable** but **short-term** record of global temperature change.

3) Here's an example of how a **temperature record** from thermometer measurements **shows** that global warming **is happening**:

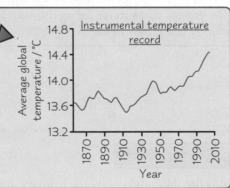

> 1) The graph on the right shows the **temperature record** from thermometer measurements.
>
> 2) Average global temperature **fluctuated** around **13.6 °C** between **1850** and **1910**.
>
> 3) It has **steadily increased** (with a couple of fluctuations) from **13.6 °C** in **1910** to around **14.4 °C** today.
>
> 4) The **general trend** of **increasing** global temperature over the last century (since 1910) is **evidence** for **global warming**.

② Dendrochronology (Tree rings)

1) **Dendrochronology** is a method for figuring out **how old** a tree is using **tree rings** (the rings formed within the trunk of a tree as it grows).

2) Most trees produce **one ring** within their trunks **every year**.

3) The **thickness** of the ring depends on the **climate** when the ring was formed — when it's **warmer** the rings are **thicker** (because the conditions for growth are better).

4) Scientists can take **cores** through **tree trunks** then **date** each ring by **counting** them **back** from when the core was taken. By looking at the **thickness** of the rings they can see what the **climate** was like **each year**.

5) Here's an example of how dendrochronology **shows** that global warming **is happening**:

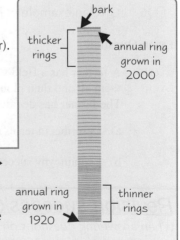

> 1) The diagram on the right shows a **core** taken from a **tree** in 2000.
>
> 2) The **most recent** rings are the **thickest** and the rings get **steadily thinner** the further in the **past** they were formed.
>
> 3) The **trend** of increasingly thicker rings from **1920** to **2000** suggests that the climate where the tree lived had become **warmer** over the **last century**.

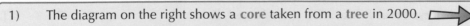

Introduction to Global Warming

③ Pollen in Peat Bogs

Pollen in peat bogs can be used to show how **temperature** has **changed** over **thousands** of years. Here's how it works:

1) **Pollen** is often **preserved** in **peat bogs** (acidic wetland areas).

2) Peat bogs accumulate in **layers** so the **age** of the preserved **pollen increases** with **depth**.

3) Scientists can take **cores** from peat bogs and extract **pollen grains** from the different aged layers. They then **identify** the **plant species** the pollen came from.

4) Only **fully grown** (mature) plant species **produce pollen**, so the samples only show the species that were **successful** at that time.

5) Scientists know the **climates** that different plant species live in **now**. When they find preserved pollen from **similar plants**, it indicates that the **climate** was **similar** when that pollen was **produced**.

6) Because plant species **vary** with **climate** the preserved pollen will **vary** as climate **changes** over time.

7) So a gradual **increase** in **pollen** from a plant species that's **more successful** in **warmer climates** would show a **rise** in **temperature** (a decrease in pollen from a plant that needs cold conditions would show the same thing).

8) Here's an example of how pollen in peat bogs can provide **evidence** for global warming events in the **past**:

> 1) The table shows data on **samples** of **pollen** taken from a **core** of a **peat bog**.
>
> 2) Between **7100** and **3100 years ago** the number of **oak tree** pollen grains **increased** from **51** grains to **253** grains.
>
> 3) This suggests that the **climate** in the area had become **better** for **oak trees** — **more** oak trees **reached maturity** and **produced pollen**.
>
> 4) Between **7100** and **3100** years ago the number of **fir tree** pollen grains in the sample **decreased** from **231** grains to **28** grains.
>
> 5) This suggests that the **climate** in the area had become **worse** for **fir trees** — **fewer** fir trees **reached maturity** and **produced pollen**.
>
> 6) Today, **oak trees** are mainly found in **temperate** (mild) regions, and **fir trees** are mainly found in **cooler** regions.
>
> 7) This suggests that the **temperature** around the peat bog **increased** over this time period — a **warming event** had occurred.

Depth of sample (metres)	Approximate age of sample (years)	Number of pollen grains in sample	
		Oak	Fir
0.5	3100	253	28
1.0	4200	194	121
1.5	5/00	138	167
2.0	7100	51	231

Practice Questions

Q1 Define global warming.

Q2 Explain why only pollen from successful plant species is preserved in peat bogs.

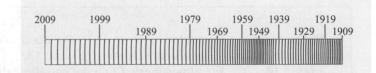

Exam Questions

Q1 The diagram on the right shows a core taken from a pine tree in 2009. Describe what the core is showing, and explain how this provides evidence for global warming. [6 marks]

Q2 How can the pollen of present-day species be used to show what the climate was like in the past? [2 marks]

I'm actually a dendrochronologist by trade — oh, you've fallen asleep...

A lot of people get global warming confused with climate change — make sure you know that global warming's just the rapid increase in global temperature over the last century. I bet you'd already thought of using a thermometer to show that global warming is happening — using tree rings and pollen isn't as obvious, but they're important sources of evidence.

Causes of Global Warming

Now you know what it is, it's time to find out what causes it...

Global Warming is Caused by Human Activity

1) The **scientific consensus** is that the recent increase in global temperature (global warming) is **caused** by **human activity**.

2) Human activity has caused global warming by **enhancing** the **greenhouse effect** — the effect of greenhouse gases absorbing outgoing **energy**, so that less is **lost** to space.

3) The greenhouse effect is **essential** to keep the planet warm, but **too much** greenhouse gas in the atmosphere means the planet **warms up**.

4) **Two** of the main greenhouse gases are CO_2 and **methane**:

Carbon dioxide (CO_2)

- **Atmospheric CO_2** concentration has **increased rapidly** since the **mid-19th century** from **280 ppm** (parts per million) to nearly **380 ppm**. The concentration had been **stable** for the previous **10 000 years**.

- CO_2 concentration is **increasing** as more **fossil fuels** like coal, oil, natural gas and petrol are **burnt**, e.g. in power stations or in cars. Burning fossil fuels **releases CO_2**.

- CO_2 concentration is also **increased** by the **destruction** of **natural sinks** (things that keep CO_2 **out** of the atmosphere by storing **carbon**). E.g. trees are a big CO_2 sink — they store the carbon as **organic compounds**. CO_2 is **released** when trees are **burnt**, or when **decomposers break down** the organic compounds and **respire** them.

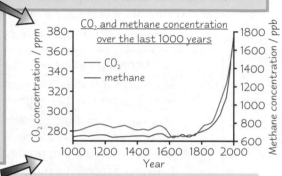

Methane (CH_4)

- **Atmospheric methane** concentration has **increased rapidly** since the **mid-19th century** from **700 ppb** (parts per billion) to **1700 ppb** in **2000**. The level had been **stable** for the previous **850 years**.

- Methane concentration is **increasing** because **more** methane is being **released** into the atmosphere, e.g. because **more fossil fuels** are being **extracted**, there's more **decaying waste** and there are **more cattle** which give off methane as a **waste gas**.

- Methane can also be released from **natural stores**, e.g. **frozen ground** (permafrost). As temperatures **increase** it's thought these stores will **thaw** and release **large amounts** of methane into the atmosphere.

An increase in **human activities** like **burning fossil fuels** (for industry and in cars), **farming** and **deforestation** has **increased** atmospheric concentrations of CO_2 and methane. This has **enhanced** the greenhouse effect and **caused** a rise in average global temperature — **global warming**.

You Need to be able to Interpret Evidence for the Causes of Global Warming

You need to be able to **interpret data** on atmospheric CO_2 concentration and temperature, and recognise **correlations** (a **relationship** between two variables) and **causal relationships** (where a change in one variable **causes** a change in another variable). Here's an example of how it's done:

1) **Describe the data:**

The **temperature fluctuated** between **1958** and **2008**, but the general trend was a **steady increase** from around **13.9 °C** to around **14.4 °C**. The atmospheric CO_2 concentration also showed a trend of **increasing** from around **315 ppm** in **1958** to around **385 ppm** in **2008**.

2) **Draw a conclusion:**

There's a **positive correlation** between the temperature and CO_2 concentration. The increasing CO_2 **concentration** could be **linked** to the increasing **temperature**. However, you **can't conclude** from this data that it's a **causal relationship** — **other factors** may have been involved, e.g. changing solar activity. **Other studies** would need to be carried out to **investigate** the effects of other factors.

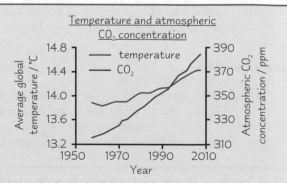

Causes of Global Warming

Some People **Disagree** About Whether **Humans** are **Causing Global Warming**

1) It's agreed that **global warming** is **happening** — there **has** been a rapid rise in global temperature over the past century.

2) It's also agreed that **human activity** is **increasing** the **atmospheric CO_2 concentration**.

3) The **scientific consensus** is that the **increase** in atmospheric CO_2 concentration **is causing** the **increase** in global temperature (i.e. humans are causing global warming).

4) But a **handful** of scientists have drawn a **different conclusion** from the **data** on atmospheric CO_2 concentration and temperature — they think that the **increase** in atmospheric CO_2 concentration **isn't** the **main cause** of the **increase** in global temperature.

5) The conclusions scientists reach can be affected by **how good** the **data** is that they're basing their conclusions on (i.e. how **reliable** it is), **how much evidence** there is for a certain theory, and also sometimes by **bias**.

6) Biased conclusions **aren't objective** — they've been **influenced** by an **opinion**, instead of being **purely** based on **scientific evidence**.

7) For example, the conclusions research scientists reach may be **biased towards** the goals of the **organisation funding** their work:

- A scientist working for an **oil company** may be more likely to say humans **aren't** causing global warming — this would help to **keep oil sales high**.

- A scientist working for a **renewable energy company** may be more likely to say humans **are** causing global warming — this would **increase sales** of energy produced from renewable sources, e.g. from wind turbines.

Trust me, humans definitely aren't causing global warming. Now, let's talk about where to build my new oil refinery.

Practice Questions

Q1 What is the greenhouse effect?

Q2 State one human activity that increases atmospheric carbon dioxide concentration.

Q3 State two human activities that increase atmospheric methane concentration.

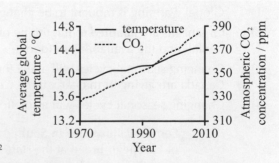

Exam Question

Q1 The graph shows the average global temperature and atmospheric CO_2 concentration from 1970 to 2008.

a) Describe the changes that the graph is showing. [4 marks]

b) Draw a conclusion about the relationship between atmospheric CO_2 concentration and temperature shown on the graph. [2 marks]

Earth not hot enough for you — spice it up with a dash of CO_2...

Another fine mess that humanity's gotten itself into — too much of things like driving, leaving TVs on standby, making big piles of rubbish and cows farting has caused global warming. I suspect you've come to the conclusion that revising global warming isn't the most fun in the world. However, I suspect that your conclusion is biased — it's so much fun...

Effects of Global Warming

The world's getting hotter, we know that much — but global warming has a couple of other tricks up its sleeve...

Global Warming Has **Different Effects**

Global warming will **directly affect plants** and **animals**. It will also change **global rainfall patterns** and the **timing of seasonal cycles**, which will also affect plants and animals:

1 *Rising Temperature*

1) An **increase** in **temperature** will affect the **metabolism** of **all** organisms:

- Normally an **increase** in **temperature** causes an **increase** in **enzyme activity**, which **speeds up** metabolic reactions.
- Enzymes have a specific **optimum temperature** — they're **most active** at this temperature.
- When temperature increases **above** the optimum temperature enzyme activity **decreases**, which **slows down** metabolic reactions.

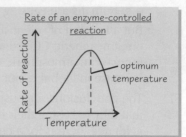

2) So an **increase** in **temperature** will mean the metabolic reactions in some organisms will **speed up**, so their **rate** of **growth** will **increase**. This also means they'll **progress** through their **life cycle faster**.

3) But the temperature will become **too high** for some organisms. Their metabolic reactions will **slow down**, so their **rate** of **growth** will **decrease**. This also means they'll **progress** through their **life cycle slower**.

4) Global warming will also affect the **distribution** of some species — all species exist where their **ideal conditions** for survival are, e.g. their ideal temperature. When these conditions **change**, they'll have to **move** to a **new area** where the conditions are better. If they **can't move** they may **die out** in that area. Also, the **range** of some species may **expand** if the conditions in previously uninhabitable areas change.

2 *Changing Rainfall Patterns*

1) Global warming will **change** global **rainfall patterns** — some areas will get **more rain**, others will get **less rain**.

2) Changing rainfall patterns will affect the **life cycles** of some organisms, e.g. ocotillo is a desert plant — it's dormant during dry periods, but after rainfall it becomes active and grows new leaves. Reduced rainfall will cause ocotillo plants to remain dormant for longer periods.

3) Changing rainfall patterns will also affect the **distribution** of some species, e.g. deserts could increase in area because of decreases in rainfall — species that aren't adapted to live in deserts will have to move to new areas or they'll die out.

3 *Seasonal Cycles*

1) Global warming is thought to be changing the **timing of the seasons**, e.g. when winter changes to spring.

2) Organisms are **adapted** to the timing of the seasons and the **changes** that happen, e.g. changes in temperature, rainfall and the availability of food.

3) Changing seasonal cycles will affect the **life cycles** of some organisms, e.g. some red squirrels in Canada are giving birth nearly three weeks earlier than usual because of an earlier availability of food.

4) Changing seasonal cycles will also affect the **distribution** of some species, for example:

1) Some **swallows** live in **South Africa** over the **winter** and fly to different parts of **Europe** to **breed** at the start of **spring** (when more food is available).

2) An **early British spring** will produce **flowers** and **insects** earlier than usual, so the swallows that migrate to Britain at the normal time will **arrive** when there **isn't** as much **food available** (there'll be **fewer insects** because the flowers will have **disappeared** earlier).

3) This will **reduce** the number of swallows that are born in Britain, and could **eventually** mean that the population of **swallows** that migrate to Britain will **die out**. The **distribution** of swallows in Europe will have **changed**.

I told you we should've come back earlier.

Effects of Global Warming

You Need to Know *How* to *Investigate* the Effect of *Temperature* on *Organisms*

Global warming will affect the **development** of **plants** and **animals**. You need to know how to **investigate** the effect of temperature on **seedling growth rate** and **brine shrimp hatch rate**:

1 Seedling Growth Rate

1) **Plant** some seedlings in **soil trays** and **measure** the **height** of each seedling.
2) Put the trays in **incubators** at **different temperatures**.
3) Make sure **all other variables** (e.g. the water content of the soil, light intensity and CO_2 concentration) are the **same** for **each tray**.
4) After a period of incubation record the **change in height** of each seedling. The **average growth rate** in each tray can be calculated in the following way:

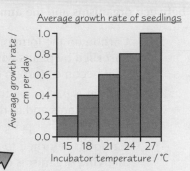
Average growth rate of seedlings

$$\frac{\text{average change in seedling height in each tray}}{\text{incubation period}}$$

5) For example, the **graph on the right** shows that as **temperature increases**, seedling **growth rate increases** — from **0.2 cm per day** at **15 °C** to **1.0 cm per day** at **27 °C**. You can **conclude** that **higher temperatures** cause **faster growth rates** (it's a **causal relationship**) because **all other variables** were **controlled**.

2 Brine Shrimp Hatch Rate

Brine shrimp are also known as Sea-Monkeys®.

1) Put an **equal number** of brine shrimp eggs in **water baths** set at **different temperatures**.
2) Make sure **all other variables** (e.g. the volume of water, the salinity of the water and O_2 concentration) are the **same** for **each water bath**.
3) The **number** of **hatched brine shrimp** in each water bath are recorded every five hours. The **hatch rate** in each water bath can be calculated in the following way:

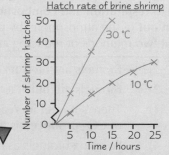
Hatch rate of brine shrimp

$$\frac{\text{number of hatched brine shrimp in each water bath}}{\text{number of hours}}$$

4) For example, the **graph on the right** shows that as **temperature increases**, brine shrimp **hatch rate increases**, e.g. at **30 °C** the initial hatch rate is **3 per hour** and at **10 °C** it's **1 per hour**. You can **conclude** that **higher temperatures** cause **faster hatch rates** (it's a **causal relationship**) because **all other variables** were **controlled**.

Practice Questions

Q1 Give one example of how changing rainfall patterns could affect the distribution of a plant or an animal.

Q2 Give one example of how the timing of the seasonal cycles could affect the life cycle of a plant or an animal.

Exam Questions

Q1 A potato tuber moth completes its life cycle faster at 21 °C than at 16 °C.
Explain why this is the case. [4 marks]

Q2 Describe how a student could investigate the effect of global warming on seedling growth rate. [5 marks]

I know what you're thinking — why do I need to know about brine shrimp...

Higher temperatures make a weekend away at an English coastal town sound more appealing, but I doubt many plants or animals will appreciate it too much. An earlier spring sounds good too — Easter eggs in February, anyone? That's if the Easter Bunny hasn't died of starvation. Thanks a lot global warming... oops, it's got all morbid and serious.

Reducing Global Warming

Since global warming will have some pretty dire consequences, some humans are having a pop at reducing it.

There are Different Ways to Reduce Atmospheric CO₂ Concentration

Increasing atmospheric CO₂ concentration is one of the **causes** of global warming (see p. 22). Scientists need to know how **carbon compounds** are **recycled** between **organisms** and the **atmosphere** so they can come up with ways to **reduce atmospheric CO₂ concentration**. The **movement** of carbon **between organisms** and the **atmosphere** is called the **carbon cycle**:

1) **Carbon** (in the form of **CO₂** from the **atmosphere**) is **absorbed** by plants when they carry out **photosynthesis** — it becomes carbon compounds in **plant tissues**.

2) Carbon is **passed on** to **animals** when they **eat** the plants and to **decomposers** when they eat **dead organic matter**.

3) Carbon is **returned** to the atmosphere as **all living organisms** carry out **respiration**, which **produces CO₂**.

4) If dead organic matter ends up in places where there **aren't any decomposers**, e.g. deep oceans or bogs, the carbon compounds can be turned into **fossil fuels** over **millions of years** (by heat and pressure).

5) The carbon in fossil fuels is **released** as **CO₂** when they're **burnt** — this is called **combustion**.

To **reduce atmospheric CO₂ concentration** either the **amount** of CO₂ **going into** the atmosphere (due to **respiration** and **combustion**) needs to be **decreased** or the **amount** of CO₂ being **taken out** of the atmosphere (by **photosynthesis**) needs to be **increased**. You need to know about two methods of reducing atmospheric CO₂ concentration:

Biofuels

1) Biofuels are **fuels** produced from **biomass** — material that **is** or **was recently living**.

2) Biofuels are **burnt** to release energy, which **produces CO₂**.

3) There's **no net increase** in atmospheric CO₂ concentration when biofuels are burnt — the amount of CO₂ **produced** is the **same** as the amount of CO₂ **taken in** when the material was **growing**.

4) So using biofuels as an **alternative** to fossil fuels **stops** the **increase** in atmospheric CO₂ concentration caused by burning fossil fuels.

Reforestation

1) Reforestation is the planting of **new trees** in **existing forests** that have been **depleted**.

2) **More trees** means **more CO₂** is **removed** from the atmosphere by **photosynthesis**.

3) CO₂ is **converted** into carbon compounds and **stored** as plant tissues in the trees. This means more carbon is **kept out** of the atmosphere, so there's **less CO₂** contributing to global warming.

People Disagree About How to Reduce Global Warming

Scientists have come up with lots of **strategies** that all **reduce global warming**. There's **debate** about which strategies are the **right** ones to use because **different people** have **different viewpoints**. Here are a few examples of why different people might support or oppose **increasing** the use of **biofuels** and **wind turbines** for energy production:

Increase the use of biofuels

- Some **farmers** might **support** this strategy — some governments **fund** the **farming** of **crops** for biofuels.
- **Drivers** might **support** this strategy — the **price** of **biofuels** is usually **lower** than **oil-based fuels**.
- **Consumers** might **oppose** this strategy — using **farmland** to grow **crops** for biofuels could cause **food shortages**.
- **Conservationists** might **oppose** this strategy — **forests** have been **cleared** to grow **crops** for biofuels.

Reducing Global Warming

Increase the use of wind turbines

- Companies that make **wind turbines** would **support** this strategy — their **sales** would **increase**.
- **Environmentalists** might **support** this strategy — wind turbines produce electricity **without increasing** atmospheric CO_2 concentration.
- **Local communities** might **oppose** this strategy — some people think wind turbines **ruin** the **landscape**.
- **Bird conservationists** might **oppose** this strategy — many **birds are killed** by **flying into** wind turbines.

Data about Global Warming can be Extrapolated to make Predictions

Data that's **already** been collected on atmospheric CO_2 concentration can be **extrapolated** — used to make **predictions** about how it will **change** in the **future**. These predictions can then be used to **model** the amount of **global warming** that might happen in the **future**. For example:

1) An **international** group of scientists called the Intergovernmental Panel on Climate Change (**IPCC**) has **extrapolated data** on atmospheric CO_2 concentration to produce a number of **emissions scenarios** — **predictions** of how human CO_2 emissions will **change** up until 2100.

2) Scenarios include:
   - Emissions continuing to **grow** as they are now ('**business as usual**').
   - Emissions **increasing** by a lot (scenario 1, maximum emissions).
   - Emissions being **controlled** by **management strategies**.
   - Emissions **not increasing** much more (scenario 5, minimum emissions).

3) They can put all these different scenarios into **global climate models** (computer models of how the climate works), to see **how much** global **temperature** will **rise** with each scenario.

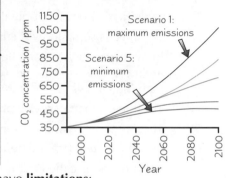

Models of future global warming based on extrapolated CO_2 concentration data have **limitations**:

1) We don't actually know how CO_2 emissions will **change**, i.e. which emissions scenario is most **accurate**.

2) We don't know exactly how much each emissions scenario will **cause** the global **temperature** to **rise by**.

3) The change in atmospheric CO_2 concentration due to **natural causes** (without human influence) **isn't known**.

4) We don't know what attempts there will be to **manage** atmospheric CO_2 concentration, or how **successful** they'll be.

In this scenario, an athlete is floored by emissions.

Practice Questions

Q1 What is reforestation?

Q2 Suggest one group of people who might oppose increasing the use of biofuels to reduce global warming.

Q3 Suggest one group of people who might support increasing the use of wind turbines to reduce global warming.

Exam Questions

Q1 Explain what biofuels are and describe how they help to reduce atmospheric CO_2 concentration. [4 marks]

Q2 Describe the limitations to models of global warming based on extrapolated CO_2 concentration data. [4 marks]

A massive, damp flannel — that's how I'd reduce global warming...

There's always talk of the next big idea to reduce global warming, but I reckon it's just hot air. Ho ho ho, sorry about that. We'll all probably have melted by the time everyone agrees on the best way to reduce global warming, but you can start doing your bit now — stop using your private jet to attend lessons and dedicate your life to planting trees.

Evolution, Natural Selection and Speciation

These pages are all about evolution and how evolution leads to new species. Unfortunately for some species, the biologists had run out of good names, e.g. Colon rectum (a type of beetle) and Aha ha (an Australian wasp). Oh dear.

Evolution *is a* Change *in* Allele Frequency

1) The complete range of **alleles** present in a **population** is called the **gene pool**.

2) **New alleles** are usually generated by **mutations** in **genes** — these are **changes** in the **base sequence** of DNA that occur during **DNA replication**.

3) How **often** an **allele occurs** in a population is called the **allele frequency**. It's usually given as a **percentage** of the total population, e.g. 35%, or a **number**, e.g. 0.35.

4) The **frequency** of an **allele** in a population **changes** over time — this is **evolution**.

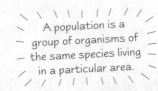
A population is a group of organisms of the same species living in a particular area.

Evolution *Occurs by* Natural Selection

1) **Individuals** within a population **vary** because they have **different alleles**.

2) This means some individuals are **better adapted** to their environment than others.

3) Individuals that have an allele that **increases** their **chance of survival** (a **beneficial** allele) are **more likely** to **survive, reproduce** and **pass on** their genes (including the beneficial allele), than individuals with different alleles.

4) This means that a **greater proportion** of the next generation **inherit** the **beneficial allele**.

5) They, in turn, are **more likely** to **survive, reproduce** and **pass on** their genes.

6) So the **frequency** of the beneficial allele **increases** from generation to generation.

7) This process is called **natural selection**.

There are different alleles due to mutations.

Speciation *is the Development of a* New Species

1) A **species** is defined as a group of **similar organisms** that can **reproduce** to give **fertile offspring**.

2) **Speciation** is the development of a **new species**.

3) It occurs when **populations** of the **same species** become **reproductively isolated** — **changes** in **allele frequencies** cause changes in **phenotype** that mean they can **no longer breed** together to produce **fertile offspring**.

Reproductive Isolation *Occurs in Many Ways*

Here are some of the ways **changes** in **phenotype prevent** two populations from **successfully breeding together**:

1) **Seasonal changes** — individuals from the same population develop different **flowering** or **mating** seasons, or become **sexually active** at **different times** of the year.

2) **Mechanical changes** — changes in **genitalia** prevent successful mating.

3) **Behavioural changes** — a group of individuals develop **courtship rituals** that **aren't attractive** to the main population.

Janice's courtship ritual was still successful in attracting mates.

A population could become **reproductively isolated** due to **geographical isolation** (see next page) or **random mutations**. Random mutations could occur **within a population**, resulting in the changes mentioned above, **preventing** members of that population breeding with other members of the species.

Evolution, Natural Selection and Speciation

Geographical Isolation and Natural Selection Lead to Reproductive Isolation

1) Geographical isolation happens when a **physical barrier divides** a population of a species — **floods**, **volcanic eruptions** and **earthquakes** can all cause barriers that isolate some individuals from the main population.

2) **Conditions** on either side of the barrier will be slightly **different**.
For example there might be a **different climate** on each side.

3) Because the environment is different on each side, **different characteristics** (phenotypes) will become **more common** due to **natural selection**:

- Because **different characteristics** will be **advantageous** on **each side**, the **allele frequencies** will **change** in **each population**, e.g. if one allele is more advantageous on one side of the barrier, the frequency of that allele on that side will **increase**.

- **Mutations** will take place **independently** in each population, also **changing** the **allele frequencies**.

- The changes in allele frequencies will lead to changes in **phenotype frequencies**, e.g. the advantageous characteristics (**phenotypes**) will become more common on that side.

4) Eventually, individuals from different populations will have changed so much that they won't be able to breed with one another to produce **fertile** offspring — they'll have become **reproductively isolated**.

5) The two groups will have become separate **species**.

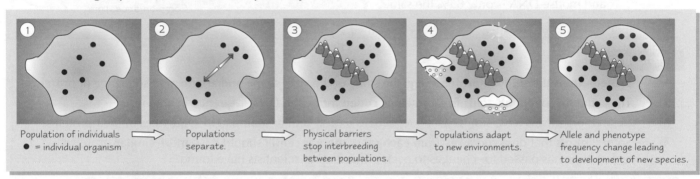

1 Population of individuals
● = individual organism

2 Populations separate.

3 Physical barriers stop interbreeding between populations.

4 Populations adapt to new environments.

5 Allele and phenotype frequency change leading to development of new species.

Practice Questions

Q1 What is a gene pool?

Q2 What term describes a change in the allele frequency within a population?

Q3 What is speciation?

Exam Question

Q1 The diagram shows an experiment conducted with fruit flies. One population was split in two and each population was fed a different food. After many generations the two populations were placed together and it was observed that they were unable to breed together.

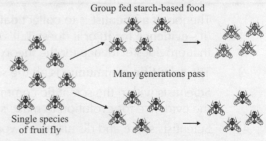

Group fed starch-based food

Many generations pass

Single species of fruit fly

Group fed maltose-based food

a) What evidence shows that speciation occurred? [1 mark]

b) Explain why the experiment resulted in speciation. [3 marks]

c) Suggest two possible reasons why members of the two populations were not able to breed together. [2 marks]

d) During the experiment, populations of fruit flies were artificially isolated. Suggest one way that populations of organisms could become isolated naturally. [1 mark]

Chess club members — self-enforced reproductive isolation...

These gags get better and better... Anyway, it's a bit of a toughie getting your head round reproductive isolation and the different ways it can occur. But don't worry, go over it a few times and you'll be laughing all the way to a top grade.

Evidence for Evolution

In science, you can't just do some research and then wave it about like it's pure fact. Oh no. A load of other scientists stick their noses in first — and then if you're lucky, your evidence will be accepted. Take evolution for example...

There's **Plenty of Evidence** to **Support Evolution**

The theory of evolution has been around for quite a long time now and there's plenty of **evidence** to support it. Fairly **new** evidence includes some from **molecular biology** — the study of **molecules** such as **DNA** and **proteins**:

DNA evidence

1) The theory of evolution suggests that all organisms have **evolved** from shared **common ancestors**.

2) Closely related species **diverged** (evolved to become different species) **more recently**.

3) Evolution is caused by **gradual changes** in the **base sequence** of organisms' DNA.

4) So, organisms that diverged away from each other more recently should have **more similar DNA**, as **less time** has passed for changes in the DNA sequence to occur. This is exactly what scientists have found.

> **Example** — Humans, chimps and mice all evolved from a common ancestor. Humans and mice diverged a **long time ago**, but humans and chimps diverged **quite recently**. The **DNA base sequence** of humans and chimps is 94% the same, but human and mouse DNA is only 85% the same.

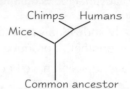

Proteomics

1) **Proteomics** is the study of **proteins**, e.g. the study of the **size**, **shape** and **amino acid sequence** of **proteins**.

2) The **sequence** of **amino acids** in a protein is **coded for** by the **DNA sequence** in a gene.

3) **Related** organisms have **similar DNA sequences** and so **similar amino acid sequences** in their proteins.

4) So organisms that diverged away from each other **more recently** should have **more similar proteins**, as **less time** has passed for changes to occur. This is what scientists have found.

The **Scientific Community Validates Evidence** About **Evolution**

1) The job of a scientist is to collect **data** and use it to **test theories** and **ideas** — the data either **supports** the theory (it's **evidence for it**) or it doesn't (it's **evidence to disprove it**). E.g. DNA and proteomic data has been collected that provides evidence for the theory of evolution.

2) The **scientific community** is all the scientists around the world, e.g. researchers, technicians and professors.

3) Scientists within the scientific community **accept** the theory of **evolution** because they've **shared** and **discussed** the evidence for evolution to make sure it's **valid** and **reliable**.

4) Scientists share and discuss their work in **three main ways**:

1 Scientific Journals

> *Examples of scientific journals include Science, Nature, the British Medical Journal and the Journal of Biological Chemistry.*

1) **Scientific journals** are **academic magazines** where scientists can publish **articles** describing their work.

2) They're used to share new **ideas**, **theories**, **experiments**, **evidence** and **conclusions**.

3) Scientific journals allow other scientists to repeat experiments and see if they get the **same results** using the **same methods**.

4) If the results are **replicated** over and over again, the scientific community can be pretty confident that the evidence collected is **reliable**.

Evidence for Evolution

2 Peer Review

1) **Before** scientists can get their work **published** in a journal it has to undergo something called the **peer review process**.

2) This is when **other scientists** who work in that area (**peers**) read and **review** the work.

Jim's peers thought this was pretty good science.

3) The peer reviewer has to **check** that the work is **valid** and that it **supports** the **conclusions**.

4) Peer review is used by the scientific community to try and make sure that any scientific evidence that's published is **valid** and that experiments are carried out to the **highest possible standards**.

3 Conferences

1) **Scientific conferences** are **meetings** that scientists attend so they can **discuss** each other's work.

2) Scientists with important or interesting results might be invited to present their work in the form of a **lecture** or **poster presentation**.

3) Other scientists can then **ask questions** and **discuss** their work with them **face to face**.

4) Conferences are valuable because they're an **easy way** for the latest theories and evidence to be **shared** and **discussed**.

Practice Questions

Q1 What is proteomics?

Q2 What is the scientific community?

Q3 Give one way a scientist might present their work at a scientific conference.

Q4 Why are conferences valuable to the scientific community?

Exam Questions

Q1 Describe and explain how the theory of evolution is supported by DNA evidence. [5 marks]

Q2 Scientific journals publish evidence for theories, such as the theory of evolution. Explain how the scientific community checks that evidence published in scientific journals is valid and reliable. [4 marks]

Peer review — checking out who's got the latest mobile phone...

Congratulations, you've finished this small but perfectly formed section. All you need to do now is keep testing yourself on the information on these two pages until you're so sick of it you want to throw the book out of the window (takes about 3 goes). Then you need to retrieve your book for the next section...

The Genetic Code and Protein Synthesis

You learnt how DNA and its mysterious cousin RNA are used to produce proteins at AS, but irritatingly you need to know it at A2 as well (with a few extra bits thrown in — unlucky).

DNA is Made of Nucleotides that Have a Sugar, a Phosphate and a Base

1) DNA is a **polynucleotide** — it's made up of many **nucleotides** that each contain a **sugar** (**deoxyribose**), a **phosphate** and a **base**.

2) There are **four** possible bases — adenine (**A**), thymine (**T**), cytosine (**C**) and guanine (**G**).

3) DNA exists as a **double-helix** — a **spiral** formed when **two polynucleotide strands** are **joined together** by **hydrogen bonding** between the bases.

4) Each base can **only join** with **one** particular partner — this is called **complementary base pairing**.

5) **A** always pairs with **T**, and **G** always pairs with **C**.

 When two strands have bases that pair up the strands are said to be complementary to each other.

DNA Contains Genes Which are Instructions for Proteins

Polypeptide is just another word for a protein.

1) Genes are **sections of DNA**. They're found on **chromosomes**.

2) Genes **code** for **proteins** (polypeptides) — they contain the **instructions** to make them.

3) Proteins are made from **amino acids**. Different proteins have a **different number** and **order** of amino acids.

4) It's the **order** of **bases** in a gene that determines the **order of amino acids** in a particular **protein**.

5) Each amino acid is coded for by a sequence of **three bases** (called a **triplet** or a **codon**) in a gene.

6) **Different sequences** of bases code for different amino acids — this is the **genetic code**.

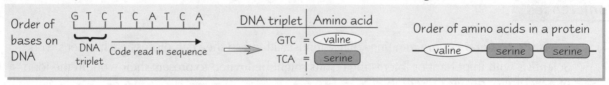

7) In the genetic code, each triplet is **read** in sequence, **separate** from the triplet **before** it and **after** it. Base triplets **don't share** their **bases** — the code is **non-overlapping**.

8) The genetic code is also **degenerate** — there are **more** possible combinations of **triplets** than there are amino acids (20 amino acids but 64 possible triplets). This means that some **amino acids** are coded for by **more than one** triplet, e.g. tyrosine can be coded for by TAT or TAC.

9) Some triplets are used to tell the cell when to **start** and **stop** production of the protein — these are called **start** and **stop codons**. They're found at the **beginning** and **end** of the gene. E.g. TAG is a stop codon.

DNA is Copied into RNA for Protein Synthesis

1) DNA molecules are found in the **nucleus** of the cell, but the organelles for protein synthesis (**ribosomes**) are found in the **cytoplasm**.

2) DNA is too large to move out of the nucleus, so a section is **copied** into **RNA**. This process is called **transcription** (see next page).

3) RNA is a **single** polynucleotide strand — it contains the sugar **ribose**, and **uracil** (**U**) replaces thymine as a base. Uracil **always pairs** with **adenine** during protein synthesis.

4) The RNA **leaves** the nucleus and joins with a **ribosome** in the cytoplasm, where it can be used to synthesise a **protein**. This process is called **translation** (see page 34).

5) There are actually two types of RNA you need to know about:

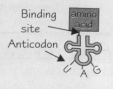

Messenger RNA (mRNA)	Transfer RNA (tRNA)
• It **carries the genetic code** from the DNA in the **nucleus** to the **cytoplasm**, where it's used to make a **protein** during **translation**. • **Three adjacent bases** are called a **codon**.	• It **carries the amino acids** that are used to make **proteins** to the **ribosomes**. • It has an **amino acid binding site** at one end and a **sequence of three bases** at the other end called an **anticodon**.

The Genetic Code and Protein Synthesis

First Stage of Protein Synthesis — Transcription

During transcription an **mRNA copy** of a gene (a section of DNA) is made in the **nucleus**:

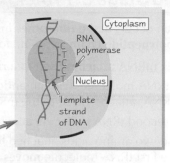

1) Transcription starts when **RNA polymerase** (an **enzyme**) **attaches** to the **DNA** double-helix at the **beginning** of a **gene**.

2) The **hydrogen bonds** between the two DNA strands in the gene **break**, **separating** the strands, and the DNA molecule **uncoils** at that point.

3) One of the strands is then used as a **template** to make an **mRNA copy**. The DNA **template strand** is also called the **antisense strand**.

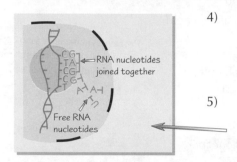

4) The RNA polymerase lines up free **RNA nucleotides** alongside the template strand. **Complementary base pairing** means that the mRNA strand ends up being a **reverse copy** of the DNA template strand (except the base **T** is replaced by **U** in **RNA**).

5) Once the RNA nucleotides have **paired up** with their **complementary bases** on the DNA strand they're **joined together**, forming an **mRNA** molecule.

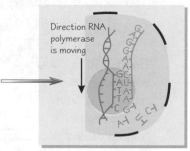

6) The RNA polymerase moves **along** the DNA, separating the strands and **assembling** the mRNA strand.

7) The **hydrogen bonds** between the uncoiled strands of DNA **re-form** once the RNA polymerase has passed by and the strands **coil back into a double-helix**.

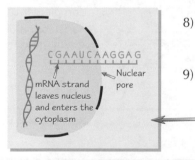

8) When RNA polymerase reaches a **stop codon**, it stops making mRNA and **detaches** from the DNA.

9) The **mRNA** moves **out** of the **nucleus** through a nuclear pore and attaches to a **ribosome** in the cytoplasm, where the next stage of protein synthesis takes place (see next page).

mRNA is Modified Before Translation

1) Genes contain sections that **don't code** for amino acids.

2) These sections of DNA are called **introns**. All the bits that **do** code for amino acids are called **exons**.

3) During transcription the introns and exons are both **copied** into mRNA.

4) The introns are then **removed** by a process called **splicing** — introns are removed and exons joined forming **mRNA** strands. This takes place in the **nucleus**.

5) The **exons** can be **joined together** in **different orders** to form **different mRNA strands**.

6) This means **more than one amino acid sequence** and so **more than one protein** can be produced from **one gene**.

7) After splicing the mRNA **leaves the nucleus** for the next stage of protein synthesis (**translation**).

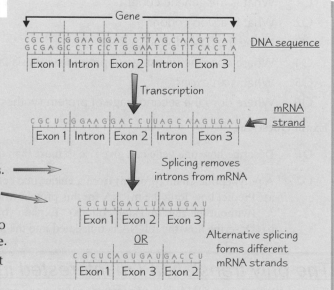

The Genetic Code and Protein Synthesis

Second Stage of Protein Synthesis — Translation

Translation occurs at the **ribosomes** in the **cytoplasm**. During **translation**, **amino acids** are **joined together** to make a **polypeptide chain** (protein), following the sequence of **codons** carried by the mRNA.

1) The **mRNA attaches** itself to a **ribosome** and **transfer RNA (tRNA)** molecules **carry amino acids** to the ribosome.

2) A tRNA molecule, with an **anticodon** that's **complementary** to the **first codon** on the mRNA, attaches itself to the mRNA by **complementary base pairing**.

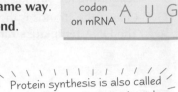

3) A second tRNA molecule attaches itself to the **next codon** on the mRNA in the **same way**.

4) The two amino acids attached to the tRNA molecules are **joined** by a **peptide bond**. The first tRNA molecule **moves away**, leaving its amino acid behind.

5) A third tRNA molecule binds to the **next codon** on the mRNA. Its amino acid **binds** to the first two and the second tRNA molecule **moves away**.

6) This process continues, producing a chain of linked amino acids (a **polypeptide chain**), until there's a **stop codon** on the mRNA molecule.

Protein synthesis is also called polypeptide synthesis as it makes a polypeptide (protein)

7) The polypeptide chain (**protein**) **moves away** from the ribosome and translation is complete.

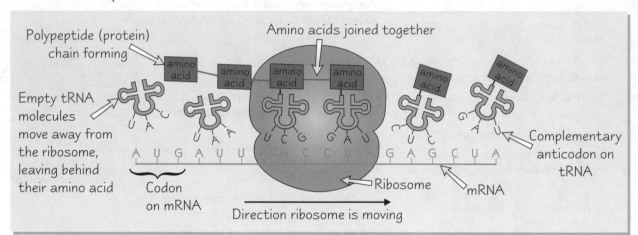

Practice Questions

Q1 What is the genetic code?

Q2 What is a stop codon?

Q3 Name the bases found in RNA.

Q4 Where does the first stage of protein synthesis take place?

Q5 What is an exon?

Q6 Where does the second stage of protein synthesis take place?

These questions are for pages 32-34.

Exam Questions

Q1 Describe how one gene can give rise to more than one protein. [5 marks]

Q2 A polypeptide chain (protein) from a eukaryotic cell is 10 amino acids long.
   a) Predict how long the mRNA for this protein would be in nucleotides (without the start and stop codons). Explain your answer. [2 marks]
   b) Describe how the mRNA is translated into the polypeptide chain. [6 marks]

The only translation I'm interested in is a translation of this page into English...

So you start off with DNA, lots of cleverness happens and bingo... you've got a protein. Only problem is you need to know the cleverness bit in quite a lot of detail. So scribble it down, recite it to yourself, explain it to your best mate or do whatever else helps you remember the joys of protein synthesis. And then think how clever you must be to know it all.

DNA Profiling

A hundred years ago they were starting to identify people using their fingerprints, but now we can use their DNA instead. DNA can also be used to figure out who's related to who... clever stuff.

A *DNA Profile* is a *Genetic Fingerprint*

1) A **DNA profile** is a **fingerprint** of an organism's DNA.

2) Everyone's DNA is **different** (except identical twins), so your DNA profile is **unique** to you.

3) DNA profiling can be used to **identify people** and to **determine genetic relationships** between plants, between animals and between humans (see pages 36-37).

4) Here's how a DNA profile is made:

1 A *DNA Sample* is *Obtained*

A **sample of DNA** is obtained from the organism you're making the DNA profile for (e.g. from **blood**, **saliva** etc.).

2 *PCR* is Used to *Amplify* the *DNA*

The **polymerase chain reaction** (PCR) is used to make **millions of copies** of **specific regions** of the DNA in just a few hours. The DNA needs to be amplified so there's **enough** to **make a DNA profile** (see the next page). PCR has **several stages** and is **repeated** over and over to make lots of copies:

1) A reaction mixture is set up that contains the **DNA sample, free nucleotides, primers** and **DNA polymerase**.
   - **Primers** are short pieces of DNA that are **complementary** to the bases at the **start** of the fragment you want.
   - **DNA polymerase** is an **enzyme** that creates new DNA strands.

2) The DNA mixture is **heated** to **95 °C** to break the **hydrogen bonds** between the two strands of DNA.

3) The mixture is then **cooled** to between **50** and **65 °C** so that the primers can **bind** (**anneal**) to the strands.

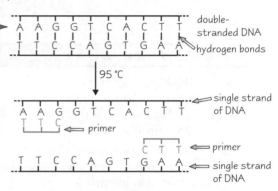

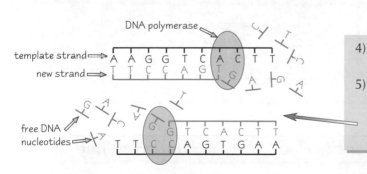

4) The reaction mixture is heated to **72 °C**, so **DNA polymerase** can **work**.

5) The DNA polymerase **lines up** free DNA nucleotides **alongside** each **template strand**. Complementary **base pairing** means **new complementary strands** are formed.

6) **Two new copies** of the fragment of DNA are formed and **one cycle** of PCR is **complete**.

7) The cycle starts again, with the mixture being heated to 95 °C and this time **all four strands** (two original and two new) are used as **templates**.

8) Each PCR cycle **doubles** the amount of DNA, e.g. **1st cycle = 2 × 2 = 4 DNA fragments, 2nd cycle = 4 × 2 = 8 DNA fragments, 3rd cycle = 8 × 2 = 16 DNA fragments**, and so on.

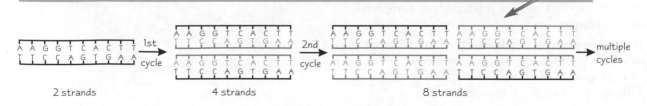

DNA Profiling

③ A *Fluorescent Tag* is *Added*

A **fluorescent tag** is added to all the DNA fragments
so they can be viewed under **UV light**.

④ *Gel Electrophoresis* is Used to *Separate* the *DNA*

Gel electrophoresis is used to **separate** out the DNA fragments according to their **length**.
Here's how it works:

1) The DNA is placed into a **well** in a slab of
gel and covered in a **buffer solution** that
conducts electricity.

2) An **electrical current** is passed through
the gel — DNA fragments are **negatively
charged**, so they **move towards** the
positive electrode at the far end of the gel.

3) **Short** DNA fragments move **faster** and
travel further through the gel, so the DNA
fragments **separate** according to **length**.

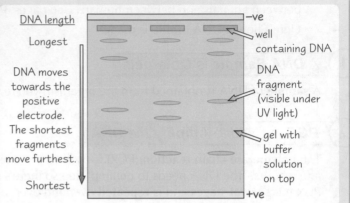

⑤ The *Gel* is *Viewed* Under *UV Light*

1) The DNA fragments
appear as **bands** under
UV light — this is the
DNA profile.

2) Two DNA profiles can be
compared — a **match**
could help **identify** a
person or determine a
genetic relationship.

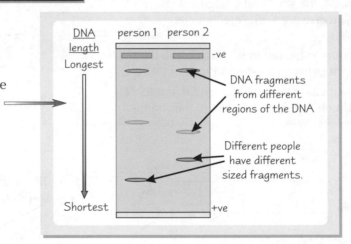

DNA Profiling can be Used to *Identify People* in *Forensic Science*

Forensic scientists use DNA profiling to **compare** samples of **DNA** collected from **crime scenes**
(e.g. DNA from **blood**, **semen**, **skin cells**, **saliva**, **hair** etc.) to samples
of DNA from **possible suspects**, to **link them** to crime scenes.

1) The **DNA** is **isolated** from all the collected samples
(from the crime scene and from the suspects).

2) Each sample is **amplified** using **PCR** (see p. 35).

3) The **PCR products** are run on an **electrophoresis
gel** and the DNA profiles produced are **compared**
to see if any match.

4) If the samples match, it **links** a **person** to the **crime
scene**. E.g. this gel shows that the DNA profile
from **suspect C matches** that from the crime scene,
linking them to the crime scene.

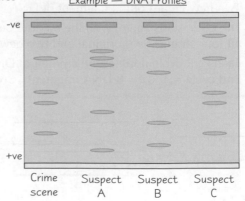

Example — DNA Profiles

DNA Profiling

DNA Profiling is Used to Determine Genetic Relationships in Humans...

1) We **inherit** our DNA from our **parents** — roughly **half** comes from **each parent**.

2) This means the **more bands** on two DNA profiles that **match**, the more **closely related** (genetically similar) those two people are.

3) For example, **paternity tests** are used to determine the **biological father** of a child by comparing DNA profiles. If lots of bands on the profile **match**, then that person is **most probably** the child's father.

Even without the hats, Cheryl and Beryl didn't think they needed a DNA profile to prove they were related...

...and in Animals and Plants

1) DNA profiling can be used on **animals** and **plants** to **prevent inbreeding**, which causes **health**, **productivity** and **reproductive problems**.

2) Inbreeding **decreases** the **gene pool** (the number of **different alleles** in a population, see p. 28), which can lead to an **increased risk** of **genetic disorders**, leading to **health problems** etc.

3) DNA profiling can be used to **identify** how **closely-related** individuals are — the **more closely-related** two individuals are, the **more similar** their DNA profiles will be (e.g. **more bands** will **match**). The **least related** individuals will be **bred together**.

Practice Questions

Q1 What is a DNA profile?

Q2 What does PCR stand for?

Q3 What is added to DNA fragments so they can be seen under UV light?

Q4 In gel electrophoresis, which electrode do DNA fragments move towards?

Q5 Why might DNA profiling be used in forensic science?

Exam Questions

Q1 PCR is used to amplify DNA in the process of making a DNA profile.
Describe and explain how to produce multiple copies of a DNA fragment using PCR. [6 marks]

Q2 The diagram on the right shows three DNA profiles — one from a child and two from possible fathers.
a) Describe how a DNA profile is made. [5 marks]
b) Which DNA profile is most likely to be from the child's father? Explain your answer. [2 marks]
c) Give another use of DNA profiling. [1 mark]

DNA profiling should give conclusive proof of who stole the biscuits...

Who would have thought that tiny pieces of DNA on a gel would be that important? Well, they are and you need to know all about them. Make sure you know the techniques used to make a DNA profile as well as its applications. And remember, it's very unlikely that two people will have the same DNA profile (except identical twins that is).

Viral and Bacterial Infections

Unfortunately, some microorganisms have got nothing better to do than cause disease. The little blighters...

Pathogens Cause Infectious Diseases

1) A **pathogen** is **any organism** that **causes disease**.

2) Diseases caused by pathogens are called **infectious diseases**.

3) Pathogenic microorganisms include **some bacteria**, **some fungi** and **all viruses**.

4) As an infectious disease **develops** in an organism it causes a **sequence of symptoms**, which **may** lead to **death**.

5) You need to know the sequence of symptoms for the diseases caused by the **human immunodeficiency virus** (HIV) and *Mycobacterium tuberculosis*:

① The Human Immunodeficiency Virus (HIV) Causes AIDS

1) The **human immunodeficiency virus** (**HIV**) infects and destroys **immune system cells**.

2) HIV infection eventually leads to **acquired immune deficiency syndrome** (**AIDS**).

3) AIDS is a condition where the immune system **deteriorates** and eventually **fails**.

4) People with HIV are classed as having AIDS when **symptoms** of their **failing immune system** start to **appear**.

5) AIDS sufferers generally develop diseases and infections that **wouldn't** cause serious problems in people with a **healthy** immune system — these are called **opportunistic infections**.

6) The length of **time** between **infection** with HIV and the **development** of AIDS **varies** between individuals but it's usually **8-10 years**. The disease then progresses through a **sequence of symptoms**:

> 1) The **initial symptoms** of AIDS include **minor infections** of mucous membranes (e.g. the inside of the nose, ears and genitals), and recurring respiratory infections. These are caused by a **lower than normal** number of **immune system cells**.
>
> *The infections become more and more serious as there are fewer and fewer immune system cells to fight them.*
>
> 2) As AIDS **progresses** the number of **immune system cells decreases** further. Patients become susceptible to **more serious infections** including chronic diarrhoea, severe bacterial infections and tuberculosis (see below).
>
> 3) During the **late stages** of AIDS, patients have a **very low number** of immune system cells and suffer from a **range of serious infections** such as toxoplasmosis of the brain (a parasite infection) and candidiasis of the respiratory system (fungal infection). It's these serious infections that kill AIDS patients, not HIV itself.

② The Bacterium Mycobacterium tuberculosis Causes Tuberculosis (TB)

1) *Mycobacterium tuberculosis* infects **phagocytes** (see p. 41) in the **lungs**.

2) It causes the lung disease **tuberculosis** (**TB**).

3) Most people **don't develop TB straight away** — their immune system **seals off** the **infected phagocytes** in structures in the lungs called **tubercles**.

4) The bacteria become **dormant** inside the tubercles and the infected person shows **no obvious symptoms**.

5) Later on, the **dormant bacteria** can become **reactivated** and **overcome** the **immune system**, causing TB.

6) **Reactivation** is **more likely** in people with **weakened immune systems**, e.g. people with **AIDS** (see above).

7) The length of **time** between the **infection** with *Mycobacterium tuberculosis* and the **development** of TB **varies** between individuals — it can be **weeks** to **years**. TB then progresses through a **sequence of symptoms**:

> 1) The **initial symptoms** of TB include **fever, general weakness** and **severe coughing**, caused by **inflammation** of the lungs.
>
>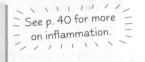
> *See p. 40 for more on inflammation.*
>
> 2) As TB **progresses** it **damages** the **lungs** and if it's left **untreated** it can cause **respiratory failure**, which can lead to **death**.
>
> 3) TB can also **spread** from the lungs to **other parts** of the body, e.g. the **brain** and **kidneys**. If it's left **untreated** it can cause **organ failure**, which can lead to **death**.

Viral and Bacterial Infections

You need to **Know** the **Structure** of **Bacteria**...

1) Bacteria are **single-celled**, **prokaryotic** microorganisms (prokaryotic means they have **no nucleus**).

2) Most bacteria are only a **few micrometers** (μm) long, e.g. the TB bacterium is about 1 μm.

3) Bacterial cells have a **plasma membrane**, **cytoplasm**, **ribosomes** and other features:

Animal and plant cells are eukaryotic.

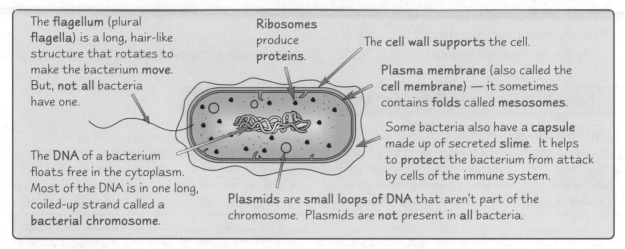

The flagellum (plural flagella) is a long, hair-like structure that rotates to make the bacterium **move**. But, **not all** bacteria have one.

The DNA of a bacterium floats free in the cytoplasm. Most of the DNA is in one long, coiled-up strand called a **bacterial chromosome**.

Ribosomes produce proteins.

Plasmids are small loops of DNA that aren't part of the chromosome. Plasmids are **not** present in **all** bacteria.

The **cell wall supports** the cell.

Plasma membrane (also called the cell membrane) — it sometimes contains **folds** called **mesosomes**.

Some bacteria also have a **capsule** made up of secreted **slime**. It helps to **protect** the bacterium from attack by cells of the immune system.

...And **Viruses**

1) Viruses are microorganisms but they're **not cells** — they're just **nucleic acids** surrounded by **protein**.

2) They're **tiny**, even **smaller** than bacteria, e.g. HIV is about 0.1 μm across.

3) **Unlike** bacteria, viruses have **no** plasma membrane, **no** cytoplasm and **no** ribosomes.

4) But they do have **nucleic acids** (like bacteria) and some other features:

Remember — DNA and RNA are nucleic acids.

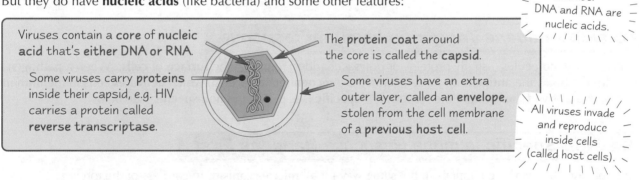

Viruses contain a **core** of nucleic acid that's **either** DNA or RNA.

Some viruses carry **proteins** inside their capsid, e.g. HIV carries a protein called **reverse transcriptase**.

The **protein coat** around the core is called the **capsid**.

Some viruses have an extra outer layer, called an **envelope**, stolen from the cell membrane of a **previous host cell**.

All viruses invade and reproduce inside cells (called host cells).

Practice Questions

Q1 What is the name of the bacterium that causes tuberculosis?

Q2 Describe two ways in which tuberculosis can cause death.

Exam Questions

Q1 People infected with HIV eventually develop AIDS.
Describe and explain the sequence of symptoms for AIDS, from infection to death. [7 marks]

Q2 All viruses are pathogens and some bacteria are pathogens.
Give three differences between the structure of bacteria and the structure of viruses. [3 marks]

My computer has a virus — I knew I shouldn't have sneezed on it...

Not the nicest of topics, but one you have to learn about — you need to know the sequence of symptoms, from infection to death, that are caused by AIDS and TB. I always find it weird just how simple a virus is. For something so basic, it can do a lot of damage to humans. I'm sure there's a moral in there somewhere, but don't worry, I won't bore you with it.

Infection and The Non-Specific Immune Response

If you're a bit worried about getting an infection, fear not — your body can stop those pesky pathogens in their tracks...

Pathogens Need to Enter the Body to Cause Disease...

Pathogens can **enter** the body via four major routes:

1) Through **cuts** in the **skin**.
2) Through the **digestive system** via **contaminated food** or **drink**.
3) Through the **respiratory system** by being **inhaled**.
4) Through other **mucosal surfaces**, e.g. the **inside** of the **nose**, **mouth** and **genitals**.

...but there are Several Barriers to Prevent Infection

Stomach acid — If you **eat** or **drink** something that contains **pathogens**, most of them will be **killed** by the **acidic** conditions of the **stomach**. However, some may **survive** and pass into the intestines where they can **invade cells** of the **gut wall** and cause disease.

Skin — Your skin acts as a **physical barrier** to pathogens. But if you **damage** your skin, **pathogens** on the surface can **enter** your **bloodstream**. The blood **clots** at the area of damage to **prevent** pathogens from entering, but some may get in **before** the clot forms.

Gut and skin flora — Your **intestines** and **skin** are **naturally covered** in billions of **harmless microorganisms** (called **flora**). They **compete** with **pathogens** for **nutrients** and **space**. This **limits** the **number** of **pathogens** living in the gut and on the skin and makes it **harder** for them to **infect** the body.

Lysozyme — **Mucosal surfaces** (e.g. eyes, mouth and nose) produce **secretions** (e.g. tears, saliva and mucus). These secretions all **contain** an **enzyme** called **lysozyme**. Lysozyme **kills bacteria** by **damaging** their **cell walls** — it makes the bacteria **burst open** (**lyse**).

Foreign Antigens Trigger an Immune Response

Antigens are **molecules** (usually proteins or polysaccharides) found on the **surface** of **cells**. When a **pathogen invades** the body, the **antigens** on its cell surface are **recognised as foreign**, which **activates** cells in the **immune system**. The body has **two types** of immune response — **specific** (see p. 42) and **non-specific** (see below).

The Non-Specific Immune Response Happens First

The non-specific response happens in the **same** way for **all microorganisms** (regardless of the foreign antigen they have) — it's **not** antigen-specific. It starts attacking the microorganisms **straight away**. You need to know about **three mechanisms** that are part of the non-specific immune response:

(1) Inflammation at the Site of Infection

The **site** where a **pathogen enters** the body (the **site of infection**) usually becomes **red**, **warm**, **swollen** and **painful** — this is called **inflammation**. Here's how it happens:

1) Immune system cells **recognise foreign antigens** on the surface of a pathogen and **release molecules** that trigger inflammation.
2) The molecules cause **vasodilation** (**widening** of the blood vessels) around the site of infection, **increasing** the **blood flow** to it.
3) The molecules also **increase** the **permeability** of the **blood vessels**.
4) The increased blood flow brings **loads** of **immune system cells** to the **site of infection** and the increased permeability allows those cells to **move out** of the blood vessels and **into** the infected tissue.
5) The immune system cells can then start to **destroy** the **pathogen**.

Trevor's throat infection triggered a small amount of inflammation.

Infection and The Non-Specific Immune Response

2) Production of **Anti-Viral Proteins** Called **Interferons**

1) When cells are **infected** with **viruses**, they produce **proteins** called **interferons**.

2) Interferons help to **prevent** viruses **spreading** to **uninfected cells**.

3) They do this in several ways:

> - They **prevent** viral **replication** by **inhibiting** the production of **viral proteins**.
> - They **activate** cells involved in the **specific** immune response (see p. 42) to **kill** infected cells.
> - They **activate** other mechanisms of the **non-specific** immune response, e.g. they **promote inflammation** to bring immune system cells to the **site of infection** (see previous page).

3) *Phagocytosis*

A **phagocyte** (e.g. a macrophage) is a type of **white blood cell** that carries out **phagocytosis** (**engulfment** of pathogens). They're found in the **blood** and in **tissues** and are the **first cells** to **respond** to a pathogen inside the body. Here's how they work:

1) A phagocyte **recognises** the **antigens** on a pathogen.

2) The cytoplasm of the phagocyte moves round the pathogen, **engulfing** it.

3) The pathogen is now contained in a **phagocytic vacuole** (a bubble) in the cytoplasm of the phagocyte.

4) A **lysosome** (an organelle that contains **digestive enzymes**) **fuses** with the phagocytic vacuole. The enzymes **break down** the pathogen.

5) The phagocyte then **presents** the pathogen's **antigens**. It sticks the antigens on its **surface** to **activate** other immune system cells (see p. 42) — so it's also called an **antigen-presenting cell**.

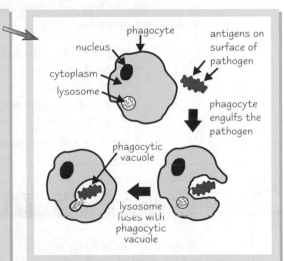

Practice Questions

Q1 State four ways in which pathogens can enter the body.

Q2 State two barriers that prevent infection.

Q3 What are antigens?

Q4 Describe two ways in which interferons prevent viruses from spreading to uninfected cells.

Exam Questions

Q1 Inflammation is part of the non-specific immune response.
Describe and explain how it occurs. [6 marks]

Q2 Describe how a phagocyte responds to an invading pathogen. [6 marks]

Studying for exams interferons with your social life...

There's a lot to remember here, but fear not, you just need to know about your body's barriers to pathogens and how the non-specific immune response works. It's also important that you know what an antigen is — it'll help you to understand how the specific immune response works. Next stop, T and B cells — and no, you can't get off...

The Specific Immune Response

Most pathogens will regret even thinking about sneaking into your body when the specific response gets going...

The **Specific Immune Response** Involves **T** and **B Cells**

The **specific immune response** is **antigen-specific** — it produces responses that are **aimed** at **specific pathogens**. It involves white blood cells called **T** and **B cells**:

1 Phagocytes **Activate T Cells**

1) A **T cell** is a type of **white blood cell**.

2) Their surface is covered with **receptors**.

3) The receptors **bind** to **antigens** presented by the phagocytes.

4) Each T cell has a **different shaped receptor** on its surface.

5) When the receptor on the surface of a T cell meets a **complementary antigen**, it binds to it — so each T cell will bind to a **different antigen**.

6) This **activates** the T cell — it **divides** and **differentiates** into **different types** of T cells that carry out **different functions**:

- **T helper cells** — **release substances** to **activate B cells** (see below), **T killer cells** and macrophages.
- **T killer cells** — **attach** to antigens on a pathogen-infected cell and **kill** the cell.
- **T memory cells** — see p. 44.

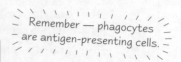

Remember — phagocytes are antigen-presenting cells.

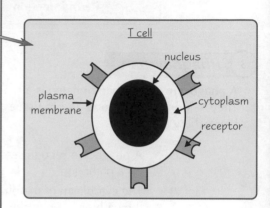

T cell

nucleus

plasma membrane

cytoplasm

receptor

A complementary antigen means it matches the shape of the receptor.

2 T Helper Cells Activate B Cells

1) **B cells** are another type of **white blood cell**.

2) They're covered with proteins called **antibodies**.

3) Antibodies **bind to antigens** to form an **antigen-antibody complex**.

4) Each B cell has a **different shaped antibody** on its surface.

5) When the antibody on the surface of a B cell meets a **complementary antigen**, it binds to it — so each B cell will bind to a **different antigen**.

6) This, together with substances **released** from the T cell, **activates** the B cell.

7) The activated B cell **divides**, by **mitosis**, into **plasma cells** (also called **B effector cells**) and **B memory cells** (see p. 44).

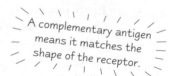

B cell

nucleus

plasma membrane

cytoplasm

antibody

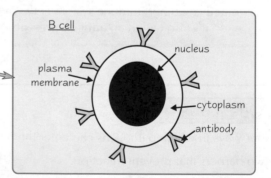

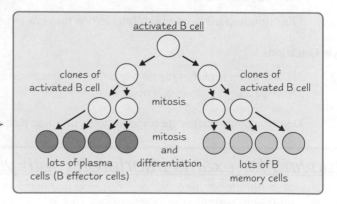

activated B cell

clones of activated B cell

clones of activated B cell

mitosis

mitosis and differentiation

lots of plasma cells (B effector cells)

lots of B memory cells

The Specific Immune Response

Plasma cells Make Antibodies to a Specific Antigen

1) **Plasma cells** are **clones** of the **B cells** (they're **identical** to the B cells).

2) They secrete **loads** of the **antibody**, specific to the antigen, into the blood.

3) These antibodies will bind to the antigens on the surface of the
pathogen to form **lots** of **antigen-antibody complexes**:

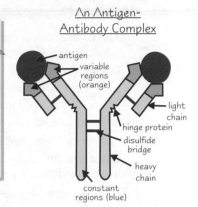

Don't forget — plasma cells are also called B effector cells.

- The **variable regions** of the antibody form the **antigen binding sites**. The **shape** of the variable region is **complementary** to a particular antigen. The variable regions **differ** between antibodies.

- The **hinge region** allows **flexibility** when the antibody binds to the antigen.

- The **constant regions** allow binding to **receptors** on **immune system cells**, e.g. phagocytes. The constant region is the **same in all** antibodies.

- **Disulfide bridges** (a type of bond) hold the polypeptide chains together.

<u>An Antigen-
Antibody Complex</u>

antigen
variable regions (orange)
light chain
hinge protein
disulfide bridge
heavy chain
constant regions (blue)

4) Antibodies **help** to **clear** an **infection** by:

1) <u>Agglutinating pathogens</u> — each antibody has **two binding sites**, so an antibody can **bind to two pathogens** at the **same time** — the pathogens become **clumped together**. Phagocytes then bind to the antibodies and phagocytose a lot of pathogens **all at once**.

2) <u>Neutralising toxins</u> — antibodies can **bind** to the **toxins** produced by pathogens. This **prevents** the toxins from **affecting human cells**, so the toxins are **neutralised** (inactivated). The toxin-antibody complexes are also phagocytosed.

3) <u>Preventing the pathogen binding to human cells</u> — when antibodies bind to the antigens on pathogens, they may **block** the cell surface **receptors** that the pathogens need to **bind to the host cells**. This means the pathogen **can't attach to** or **infect** the host cells.

<u>Agglutination</u>

antibody
pathogen antigen

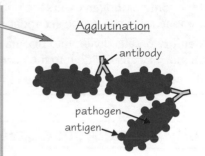

Practice Questions

Q1 What structures are found on the surface of T cells?

Q2 Briefly describe how a B cell is activated.

Q3 What cells do activated B cells divide into?

Exam Questions

Q1 a) Describe how a T cell is activated. [2 marks]

b) Name the cells that activated T cells divide and differentiate into. [3 marks]

Q2 Describe the function of antibodies. [3 marks]

The student-revision complex — only present the night before an exam...

Memory cells are still types of B and T cells. Luckily for you, you get to learn all about them on the next page. Basically they're the elephants of the immune system — they have a really long memory. They stick around in the body waiting for the same pathogen to strike again, and if it does, these cells can immediately get to work destroying it. Ha ha (evil laugh).

Developing Immunity

Your immune system has a memory — it's handy for fighting infections, not for remembering biology notes...

The Production of **Memory Cells** Gives **Immunity**

1) When a **pathogen** enters the body for the **first time** the **antigens** on its surface **activate** the **immune system**. This is called the **primary response**.

2) The primary response is **slow** because there **aren't many B cells** that can make the antibody needed to bind to the antigen.

3) Eventually the body will produce **enough** of the right antibody to overcome the infection. Meanwhile the infected person will show **symptoms** of the disease.

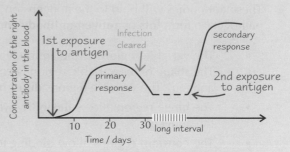

4) After being exposed to an antigen, both T and B cells produce **memory cells**. These memory cells **remain in the body** for a **long** time. Memory T cells remember the **specific antigen** and will recognise it a second time round. Memory B cells record the specific **antibodies** needed to bind to the antigen.

5) The person is now **immune** — their immune system has the **ability** to respond **quickly** to a second infection.

6) If the **same pathogen** enters the body again, the immune system will produce a **quicker**, **stronger** immune response — the **secondary response**.

7) **Memory B cells** divide into **plasma cells** that produce the right antibody to the antigen. **Memory T cells** divide into the **correct type** of T cells to kill the cell carrying the antigen.

8) The secondary response often gets rid of the pathogen **before** you begin to show any **symptoms**.

Immunity can be *Active* or *Passive*

ACTIVE IMMUNITY

This is the type of immunity you get when **your immune system makes its own antibodies** after being **stimulated** by an **antigen**. There are **two** different types of active immunity:

1) **Natural** — this is when you become immune after **catching a disease**.

2) **Artificial** — this is when you become immune after you've been given a **vaccine** containing a harmless dose of antigen (see below).

Active immunity gives you long-term protection, but it takes a while for the protection to develop.

PASSIVE IMMUNITY

This is the type of immunity you get from being **given antibodies made by a different organism** — your immune system **doesn't** produce any antibodies of its own. Again, there are **two** types:

1) **Natural** — this is when a **baby** becomes immune due to the antibodies it receives from its **mother**, through the **placenta** and in **breast milk**.

2) **Artificial** — this is when you become immune after being **injected** with **antibodies**, e.g. if you contract tetanus you can be injected with antibodies against the tetanus toxin.

Passive immunity gives you short-term protection, but the protection is immediate.

Vaccines Give You *Immunity Without Getting* the *Disease*

1) While your B cells are busy **dividing** to build up their numbers to deal with a pathogen (i.e. the **primary response** — see above), you **suffer** from the disease. **Vaccination** can help avoid this.

2) Vaccines **contain antigens** that cause your body to **produce memory cells** against a particular pathogen, **without** the pathogen **causing disease**. This means you become **immune** without getting any **symptoms**.

3) Some vaccines contain **many different antigens** to protect against **different strains** of pathogens. Different strains of pathogens are created by **antigenic variation** (see next page).

Developing Immunity

Pathogens Evolve Mechanisms to Evade the Immune System

1) Over **millions** of **years** vertebrates (e.g. humans) have **evolved** better and better **immune systems** — ones that **fight** a **greater variety** of pathogens in lots of **different ways**.

2) At the same time, **pathogens** have **evolved** better and better ways to **evade** (**avoid**) the immune systems of their **hosts** (the **organisms** that they **infect**).

3) This struggle between **pathogens** and their **hosts** to outdo each other is known as an **evolutionary race**.

4) An evolutionary race is **similar** to an **arms race** — where **two countries** constantly **develop better weapons** in an attempt to **overpower** each other.

5) **Evidence** to **support** the **theory** of an **evolutionary race** comes from the **evasion mechanisms** that pathogens have **developed**. For example:

HIV's evasion mechanisms

- HIV **kills** the **immune systems cells** that it **infects**. This **reduces** the overall **number** of immune system cells in the body, which **reduces** the **chance** of HIV being **detected**.

- HIV has a **high rate** of **mutation** in the **genes** that code for **antigen proteins**. The mutations **change** the **structure** of the antigens and this forms **new strains** of the virus — this process is called **antigenic variation**. The **memory cells** produced for **one strain** of HIV **won't recognise** other strains with **different antigens**, so the immune system has to produce a **primary response** against **each new strain**.

- HIV **disrupts antigen presentation** in infected cells. This **prevents** immune system cells from **recognising** and **killing** the **infected cells**.

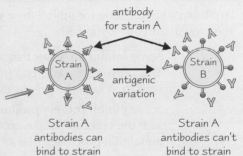

antibody for strain A

Strain A → antigenic variation → Strain B

Strain A antibodies can bind to strain A antigens

Strain A antibodies can't bind to strain B antigens

Mycobacterium tuberculosis' evasion mechanisms

- When *M. tuberculosis* bacteria **infect** the **lungs** they're **engulfed** by **phagocytes**. Here, they **produce substances** that **prevent** the **lysosome fusing** with the **phagocytic vacuole** (see p. 41). This means the bacteria **aren't broken down** and they can **multiply undetected** inside phagocytes.

- This bacterium also **disrupts antigen presentation** in infected cells, which **prevents** immune system cells from **recognising** and **killing** the **infected phagocytes**.

Practice Questions

Q1 Describe the role of memory B cells in the secondary response.
Q2 Describe the role of memory T cells in the secondary response.
Q3 State two differences between active and passive immunity.
Q4 Describe how a vaccine gives immunity to a pathogen.
Q5 Describe one way in which *Mycobacterium tuberculosis* evades the immune system.

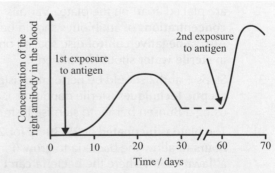

Concentration of the right antibody in the blood

1st exposure to antigen

2nd exposure to antigen

Time / days

Exam Questions

Q1 The graph shows the concentration of antibody in a person's bloodstream following two separate exposures to the same antigen. Describe and explain the changes in the concentration of antibody. [10 marks]

Q2 Describe and explain three mechanisms that HIV has evolved to evade immune systems. [6 marks]

Even if you could change your antigens, you can't evade your exams...

Who would've thought there were so many ways to become immune to a microorganism. I don't like the sound of being injected with antibodies (I'm not keen on needles), but if it stops me dying from something like tetanus, then I'll give it a go. We're in an evolutionary race with loads of nasty microbes at the minute... let's hope we're the first past the post...

Antibiotics

You've probably taken antibiotics at some point. Now you get to learn about how they work — the fun never ends...

Antibiotics Kill or Prevent the Growth of Microorganisms

1) **Antibiotics** are **chemicals** that **kill** or **inhibit** the **growth** of microorganisms.

2) There are **two different types** of antibiotics:

> - **Bacteriocidal** antibiotics **kill** bacteria.
> - **Bacteriostatic** antibiotics **prevent** bacteria **growing**.

3) Antibiotics are **used** by humans as **drugs** to **treat bacterial infections**.

Antibiotics Work by Inhibiting Bacterial Metabolism

Antibiotics kill bacteria (or inhibit their growth) because they **interfere** with **metabolic reactions** that are **crucial** for the growth and life of the cell:

1) Some inhibit enzymes that are needed to make the chemical **bonds** in bacterial **cell walls**. This prevents the bacteria from **growing** properly. It can also lead to **cell death** — the weakened cell wall can't take the **pressure** as water moves into the cell by **osmosis**. This can cause the cell to **burst**.

2) Some **inhibit protein production** by binding to bacterial **ribosomes**. All **enzymes** are proteins, so if the cell can't make proteins, it can't make enzymes. This means it can't carry out important **metabolic processes** that are needed for growth and development.

Bacterial cells are **different** from **mammalian** cells (e.g. human cells) — mammalian cells are **eukaryotic**, they **don't** have **cell walls**, they have **different enzymes** and they have different, **larger ribosomes**. This means antibiotics can be designed to **only target** the **bacterial cells**, so they don't damage mammalian cells. **Viruses don't** have their **own** enzymes and ribosomes — they use the ones in the host's cell, so antibiotics **don't affect them**.

You Can Test the Effect of Different Antibiotics on Bacteria

There are lots of reasons why you might want to **test** the **effect** of **different antibiotics** on **different strains** of bacteria. For example, doctors need to find out **which** antibiotics will **treat** a **patient's bacterial** infection.

Here's one way to do it:

1) The bacteria to be tested are **spread** onto an **agar plate**.

2) Paper discs **soaked** with **antibiotics** are placed apart on the plate. Various **concentrations** of antibiotics should be used. Also, a **negative control** disc soaked only in **sterile water** should be added.

3) Steps 1 and 2 should be performed using **aseptic techniques** (sterile conditions), e.g. using a Bunsen burner to sterilise instruments.

4) The plate is **incubated** at **25-30 °C** for **24-36 hours** to allow the bacteria to **grow** (forming a 'lawn'). Anywhere the bacteria **can't grow** can be seen as a **clear patch** in the lawn of bacteria. This is called an **inhibition zone**.

5) The size of an **inhibition zone** tells you how well an antibiotic works. The **larger** the zone, the **more** the bacteria were inhibited from growing.

6) Bacteria that are **unaffected** by antibiotics are said to be **antibiotic resistant**.

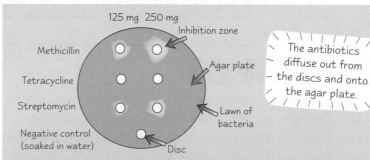

The antibiotics diffuse out from the discs and onto the agar plate.

This diagram shows an agar plate with **methicillin**, **tetracycline** and **streptomycin** discs **after** it has been **incubated**.

- The **tetracycline** discs have **no** inhibition zones, so the bacteria are **resistant** to tetracycline up to 250 mg.

- The **streptomycin** discs have **small** inhibition zones, with the zone at 250 mg slightly larger than the one at 125 mg. So streptomycin has **some effect** on the bacteria.

- The **methicillin** discs have the **largest** inhibition zones, especially at 250 mg, so methicillin has the **strongest effect** on these bacteria.

Antibiotics

Hospital Acquired Infections (HAIs) can be Transmitted by Poor Hygiene

1) **Hospital acquired infections** (HAIs) are **infections** that are **caught** while a patient is being treated **in hospital**.

2) HAIs are **transmitted** by poor hygiene, such as:

- Hospital **staff** and **visitors not washing** their **hands** before and after visiting a patient.
- **Coughs** and **sneezes not** being **contained**, e.g. in a tissue.
- **Equipment** (e.g. beds or surgical instruments) and **surfaces not** being **disinfected** after they're used.

3) People are **more likely** to catch infections in hospital because many patients are ill, so have **weakened immune systems**, and they're **around** other **ill people**.

4) **Codes of practice** have been developed to **prevent** and **control HAIs**. They include:

- Hospital **staff** and **visitors** should be **encouraged** to **wash** their **hands**, **before** and **after** they've been with a patient.
- **Equipment** and **surfaces** should be **disinfected** after they're used.
- **People with HAIs** should be **moved** to an **isolation ward** so they're **less likely** to **transmit** the infection to **other patients**.

Despite his protests, Huxley wasn't allowed in the ward with his mucky trotters.

Some HAIs are Antibiotic-Resistant

1) Some **HAIs** are caused by bacteria that are **resistant** to **antibiotics**, e.g. MRSA.

2) These HAIs are **difficult** to **treat** because antibiotics **don't** get rid of the infection. This means these HAIs can lead to **serious health problems** or even **death**.

3) Infections caused by antibiotic-resistant bacteria are **more common** in **hospitals** because **more** antibiotics are used there, so bacteria in hospitals are **more likely** to have **evolved resistance** against them.

4) **Codes of practice** have also been developed to **prevent** and **control** HAIs caused by antibiotic-resistant bacteria:

- Doctors **shouldn't** prescribe antibiotics for **minor** bacterial infections or **viral** infections.
- Doctors **shouldn't** prescribe antibiotics to **prevent** infections.
- Doctors **should** use **narrow-spectrum antibiotics** (which only affect a specific bacterium) if possible, e.g. when the **strain** of bacteria the person has is **identified**.
- Doctors **should rotate** the **use** of **different** antibiotics.
- Patients **should** take **all** of the antibiotics that they're **prescribed** so infections are **fully cleared**.

These codes reduce the likelihood that bacteria will evolve antibiotic resistance.

Practice Questions

Q1 What are bacteriostatic antibiotics?

Q2 Name two processes in a bacterial cell that antibiotics can inhibit.

Q3 What is a hospital acquired infection?

Exam Questions

Negative control (N)
Penicillin 125 mg (A)
Amoxicillin 125 mg (B)
Erythromycin 125 mg (C)
Streptomycin 125 mg (D)

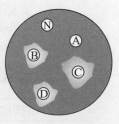

Q1 The agar plate on the right shows the effects of different antibiotics on a strain of bacteria.
  a) Describe how the plate could have been prepared. [4 marks]
  b) Which antibiotic is most effective against this strain of bacterium? Explain your answer. [2 marks]

Q2 Describe one way in which poor hygiene can cause HAIs, and a code of practice designed to prevent it. [2 marks]

The Market Research Society of Australia — not a deadly bacterium...

It's just typical, scientists discover all these fancy chemicals that get rid of bacteria and then some of the swines decide they won't be affected by them. Spoilsports. So, if you ever visit a hospital, make sure you use the alcohol gel to clean your hands. Not only will you be preventing those pesky infections, but you'll experience a cooling sensation like no other. Lovely...

Microbial Decomposition and Time of Death

Wowsers, this is pretty morbid stuff. Make sure you read these pages before you've had your lunch...

Microorganisms Decompose Organic Matter

1) Microorganisms, e.g. **bacteria** and **fungi**, are an important part of the **carbon cycle** (see p. 26).

2) When plants and animals die, microorganisms on and in them **secrete enzymes** that decompose the **dead organic matter** into **small molecules** that they can **respire**.

3) When the microorganisms respire these small molecules, **methane** and CO_2 are released — this **recycles** carbon back into the atmosphere.

Scientists can Estimate the Time of Death of a Body

Police and **forensic scientists** often need to establish a body's **time of death** (TOD). This can give them a lot of information about the **circumstances** of the death, e.g. if they know when someone died they might be able to figure out who was present. The TOD can be established by looking at **several different factors** together — on their **own** these factors **aren't accurate** enough to give a reliable time of death. The **five** factors you need to know about are:

1) Body Temperature

1) All mammals **produce heat** from metabolic **reactions** like **respiration**, e.g. the **human body** has an internal temperature of around **37 °C**.

2) From the TOD the metabolic reactions **slow down** and eventually **stop**, causing **body temperature** to **fall** until it **equals** the temperature of its **surroundings** — this process is called *algor mortis*.

3) Forensic scientists know that **human bodies** cool at a rate of around **1.5 °C** to **2.0 °C per hour**, so from the temperature of a dead body they can **work out** the approximate TOD. E.g. a dead body with a temperature of **35 °C** might have been **dead** for about an **hour**.

4) Conditions such as **air temperature**, **clothing** and **body weight** can **affect** the **cooling rate** of a body. E.g. the cooling rate of a **clothed body** will be **slower** than one without clothing, because it's **insulated**.

2) Degree of Muscle Contraction

About **4-6 hours** after death, the **muscles** in a dead body **start** to **contract** and become **stiff** — this is called *rigor mortis*:

1) *Rigor mortis* begins when **muscle cells** become **deprived** of **oxygen**.

2) **Respiration** still takes place in the muscle cells, but it's **anaerobic**, which causes a build-up of **lactic acid** in the muscle.

3) The **pH** of the cells **decreases** due to the lactic acid, **inhibiting enzymes** that produce ATP.

4) **No ATP** means the **bonds** between the **myosin** and **actin** in the muscle cells (see p. 52) become **fixed** and the body **stiffens**.

It usually takes around 12-18 hours after the TOD for every muscle in the body to contract.

Smaller muscles in the head **contract first**, with **larger muscles** in the lower body being the **last** to contract. *Rigor mortis* is affected by **degree of muscle development** and **temperature**. E.g. *rigor mortis* occurs **more quickly** at **higher** temperatures because the chemical reactions in the body are **faster**.

Rigor mortis wears off around 24-36 hours after the TOD.

3) Forensic Entomology

1) When somebody dies the body is quickly **colonised** by a **variety** of **different insects** — the study of this is called **forensic entomology**.

2) TOD can be estimated by identifying the **type of insect** present on the body — e.g. **flies** are often the **first** insects to appear, usually a **few hours** after death. Other insects, like **beetles**, colonise a body at **later** stages.

3) TOD can also be estimated by identifying the **stage of lifecycle** the insect is in — e.g. **blowfly larvae hatch** from eggs about **24 hours** after they're **laid**. If **only** blowfly **eggs** are found on a body you could estimate that the TOD was **no more** than **24 hours ago**.

4) Different conditions will **affect** an insect's **lifecycle**, such as **drugs**, **humidity**, **oxygen** and **temperature**. E.g. the **higher** the temperature, the **faster** the **metabolic rate** and the **shorter** the **lifecycle**.

Microbial Decomposition and Time of Death

4) Extent of Decomposition

1) **Immediately** after death **bacteria** and **enzymes** begin to **decompose** the **body**.

2) Forensic scientists can use the **extent** of decomposition to establish a **TOD**:

Approximate time since TOD	Extent of decomposition
Hours to a few days	Cells and tissues are being broken down by the body's own enzymes and bacteria that were present before death. The skin on the body begins to turn a greenish colour.
A few days to a few weeks	Microorganisms decompose tissues and organs. This produces gases (e.g. methane), which cause the body to become bloated. The skin begins to blister and fall off.
A few weeks	Tissues begin to liquefy and seep out into the area around the body.
A few months to a few years	Only a skeleton remains.
Decades to centuries	The skeleton begins to disintegrate until there's nothing left of the body.

3) Different conditions **affect** the **rate** of decomposition, such as **temperature** and **oxygen availability**. E.g. **aerobic** microorganisms need **oxygen**, so decomposition could be **slower** if there's a **lack** of oxygen.

5) Stage of Succession

1) The **types of organism** found in a dead body **change over time**, going through a number of **stages** — this is called **succession**.

2) Forensic scientists can establish a **TOD** from the particular **stage of succession** that the body's in.

3) If a dead body is left to decompose **above ground** succession will usually follow these stages:

- **Immediately after** the TOD conditions in a dead body are **most favourable** for **bacteria**.
- As bacteria **decompose tissues**, conditions in a dead body become favourable for **flies** and their **larvae**.
- When fly larvae **feed** on a dead body they make conditions favourable for **beetles**, so beetles move in.
- As a dead **body dries out** conditions become **less favourable** for **flies** — they leave the body. Beetles **remain** as they can decompose **dry tissue**.
- When **no tissues** remain, conditions are **no longer favourable** for **most organisms**.

4) Succession in a dead body is **similar** to plant succession (see p. 18) — the **only difference** is that most of the **early insects** (e.g. beetles) **remain** on the body as other insects colonise it.

5) The stage of succession of a dead body (and so the type of organism that's present) is affected by many things including the **location** of the body, such as above ground, under ground, in water or sealed away. E.g. a body that's been **sealed away won't** be **colonised** by any species of insect.

Practice Questions

Q1 Describe how microorganisms recycle carbon from dead plants and animals.

Q2 What is forensic entomology?

Exam Question

Q1 A human body with a temperature of 29 °C was found at 22:45. *Rigor mortis* was only present in the face and shoulders. There was no visible decomposition of the body. Blowfly eggs, but no larvae, were found on the body. For each piece of evidence above estimate the person's time of death and explain your answer. [8 marks]

CSI: Cumbria — it doesn't really have the same ring to it, does it...

Well, that was just lovely. Admittedly, I found it all rather interesting 'cause I like to watch repeats of Quincy on the telly. The main aim of all this grim stuff is to estimate the time of death of a body — remember that it's only an estimate, and lots of things (especially temperature) affect a dead body. Maybe if I pickle myself now, I'll never decompose. Hum...

Muscles and Movement

Muscles are pretty darn useful — they contract so you can move. But before we get into the gritty detail of exactly how they contract, you need to know a bit more about them and how they're involved in movement...

Movement *Involves* Skeletal Muscles, Tendons, Ligaments *and* Joints

1) **Skeletal muscle** is the type of muscle you use to **move**, e.g. the biceps and triceps move the lower arm.

2) Skeletal **muscles** are **attached** to **bones** by **tendons**.

3) **Ligaments attach bones** to **other bones**, to hold them together.

4) Skeletal muscles **contract** and **relax** to **move bones** at a **joint**.
To understand how this works it's best to look at an example:

- The bones of your **lower arm** are attached to a **biceps** muscle and a **triceps** muscle by **tendons**.
- The biceps and triceps **work together** to move your arm — as one **contracts**, the other **relaxes**:

When your **biceps contracts** your **triceps relaxes**. This pulls the bone so your **arm bends** (**flexes**) at the elbow. A muscle that **bends** a joint when it contracts is called a **flexor**.

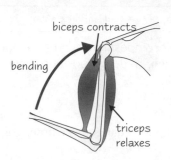

biceps contracts

bending

triceps relaxes

When your **triceps contracts** your **biceps relaxes**. This pulls the bone so your **arm straightens** (**extends**) at the **elbow**. A muscle that **straightens** a joint when it contracts is called an **extensor**.

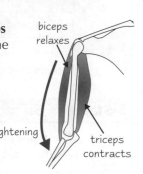

biceps relaxes

straightening

triceps contracts

- Muscles that work together to move a bone are called **antagonistic pairs**.

Muscles work in pairs because they can only pull (when they contract) — they can't push.

Skeletal Muscle *is made up of* Long Muscle Fibres

1) Skeletal muscle is made up of **large bundles** of **long cells**, called **muscle fibres**.

2) The cell membrane of muscle fibre cells is called the **sarcolemma**.

3) Bits of the sarcolemma **fold inwards** across the muscle fibre and stick into the **sarcoplasm** (a muscle cell's cytoplasm). These folds are called **transverse (T) tubules** and they help to **spread electrical impulses** throughout the sarcoplasm so they **reach** all parts of the **muscle fibre**.

4) A network of **internal membranes** called the **sarcoplasmic reticulum** runs through the sarcoplasm. The sarcoplasmic reticulum **stores** and **releases calcium ions** that are needed for muscle contraction (see p. 52).

5) Muscle fibres have lots of **mitochondria** to **provide** the **ATP** that's needed for **muscle contraction**.

6) They are **multinucleate** (contain many nuclei).

7) Muscle fibres have lots of **long, cylindrical organelles** called **myofibrils**. They're made up of proteins and are **highly specialised** for **contraction**.

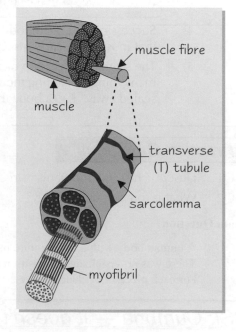

muscle fibre

muscle

transverse (T) tubule

sarcolemma

myofibril

Muscles and Movement

Myofibrils Contain Thick Myosin Filaments and Thin Actin Filaments

1) Myofibrils contain bundles of **thick** and **thin myofilaments** that **move past each other** to make muscles **contract**.
   - **Thick myofilaments** are made of the protein **myosin**.
   - **Thin myofilaments** are made of the protein **actin**.

2) If you look at a **myofibril** under an **electron microscope**, you'll see a pattern of alternating **dark** and **light bands**:
   - **Dark** bands contain the **thick myosin filaments** and some overlapping thin actin filaments — these are called **A-bands**.
   - **Light** bands contain **thin actin filaments** only — these are called **I-bands**.

3) A myofibril is made up of many short units called **sarcomeres**.

4) The **ends** of each **sarcomere** are marked with a **Z-line**.

5) In the **middle** of each sarcomere is an **M-line**. The M-line is the **middle** of the **myosin** filaments.

6) **Around** the M-line is the **H-zone**. The H-zone **only** contains **myosin** filaments.

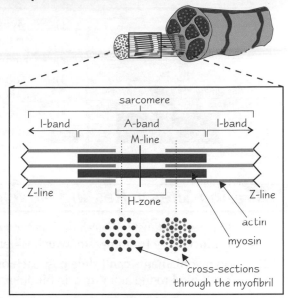

Muscle Contraction is Explained by the Sliding Filament Theory

1) **Myosin** and **actin** filaments **slide** over one another to make the **sarcomeres contract** — the myofilaments themselves **don't** contract.

2) The **simultaneous contraction** of lots of **sarcomeres** means the **myofibrils** and **muscle fibres contract**.

3) Sarcomeres return to their **original length** as the muscle **relaxes**.

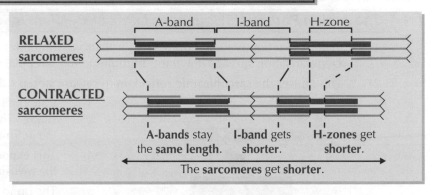

A-bands stay the **same length**. **I-band** gets **shorter**. **H-zones** get **shorter**.

The **sarcomeres** get **shorter**.

Practice Questions

Q1 What is skeletal muscle?

Q2 What are transverse (T) tubules?

Q3 Name the two proteins that make up myofibrils.

Figure 1

Exam Questions

Q1 Describe how myofilaments, muscle fibres, myofibrils and muscles are related to each other. [3 marks]

Q2 When walking, your quadriceps (muscles at the front of the thigh) contract to straighten the leg, whilst your hamstrings (muscles at the back of the thigh) relax. Then your hamstrings contract to bend the leg, whilst the quadriceps relax.
   a) State which of these muscles are the extensors and which are the flexors. [2 marks]
   b) What are muscles that work together to move a bone called? [1 mark]

Q3 A muscle myofibril was examined under an electron microscope and a sketch was drawn (Figure 1 above). What are the correct names for labels A, B and C? [3 marks]

Sarcomere — a French mother with a dry sense of humour...

Muscles involved in movement work in antagonistic pairs — one contracts as the other relaxes and vice versa. So let's practise — lower arm up, lower arm down, lower arm up... okay, that's enough exercise for one day. Anyway, you need to get on with learning all the similar-sounding names on these pages and what they mean, like myofilament and myofibril.

Muscle Contraction

Brace yourself — here comes the detail of muscle contraction...

Myosin Filaments Have Globular Heads and Binding Sites

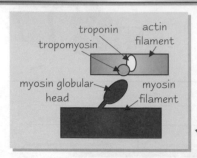

1) **Myosin filaments** have **globular heads** that are **hinged**, so they can move **back** and **forth**.

2) Each myosin head has a **binding site** for **actin** and a **binding site** for **ATP**.

3) **Actin filaments** have **binding sites** for **myosin heads**, called **actin-myosin** binding sites.

4) Two other **proteins** called **tropomyosin** and **troponin** are found between actin filaments. These proteins are **attached** to **each other** and they **help** myofilaments **move** past each other.

Binding Sites in Resting Muscles are Blocked by Tropomyosin

1) In a **resting** (unstimulated) muscle the **actin-myosin binding site** is **blocked** by **tropomyosin**, which is held in place by **troponin**.

2) So **myofilaments can't slide** past each other because the **myosin heads can't bind** to the actin-myosin binding site on the actin filaments.

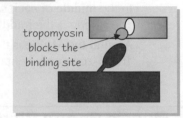

Muscle Contraction is Triggered by an Action Potential

1) The Action Potential Triggers an Influx of Calcium Ions

1) When an action potential from a motor neurone **stimulates** a muscle cell, it **depolarises** the **sarcolemma**. Depolarisation **spreads** down the **T-tubules** to the **sarcoplasmic reticulum** (see p. 50).

2) This causes the **sarcoplasmic reticulum** to **release** stored **calcium ions** (Ca^{2+}) into the **sarcoplasm**.

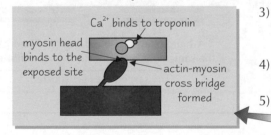

3) Calcium ions **bind** to **troponin**, causing it to **change shape**. This **pulls** the attached **tropomyosin out** of the **actin-myosin binding site** on the actin filament.

4) This **exposes** the **binding site**, which allows the **myosin head** to **bind**.

5) The bond formed when a **myosin head** binds to an **actin filament** is called an **actin-myosin cross bridge**.

2) ATP Provides the Energy Needed to Move the Myosin Head...

1) **Calcium** ions also **activate** the enzyme **ATPase**, which **breaks down ATP** (into ADP + P$_i$) to **provide** the **energy** needed for muscle contraction.

2) The **energy** released from ATP **moves** the **myosin head**, which **pulls** the **actin filament** along in a kind of **rowing action**.

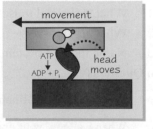

3) ...and to Break the Cross Bridge

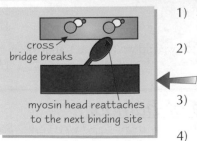

1) **ATP** also provides the **energy** to **break** the **actin-myosin cross bridge**, so the **myosin head detaches** from the actin filament **after** it's moved.

2) The **myosin head** then **reattaches** to a **different binding site** further along the actin filament. A **new actin-myosin cross bridge** is formed and the **cycle** is **repeated** (attach, move, detach, reattach to new binding site...).

3) **Many** cross bridges **form** and **break** very **rapidly**, pulling the actin filament along — which **shortens** the sarcomere, causing the **muscle** to **contract**.

4) The cycle will **continue** as long as **calcium ions** are **present** and **bound** to **troponin**.

Muscle Contraction

When **Excitation Stops**, **Calcium Ions Leave** *Troponin Molecules*

1) When the muscle **stops** being **stimulated**, **calcium ions leave** their **binding sites** on the **troponin** molecules and are moved by **active transport** back into the **sarcoplasmic reticulum** (this needs **ATP** too).

2) The **troponin** molecules return to their **original shape**, pulling the attached **tropomyosin** molecules with them. This means the **tropomyosin** molecules **block** the actin-myosin **binding sites** again.

3) Muscles **aren't contracted** because **no myosin heads** are **attached** to **actin** filaments (so there are no actin-myosin cross bridges).

4) The **actin** filaments **slide back** to their **relaxed** position, which **lengthens** the **sarcomere**.

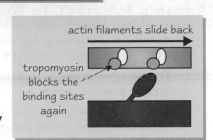

actin filaments slide back

tropomyosin blocks the binding sites again

Skeletal Muscles are Made of *Slow* and *Fast Twitch Muscle Fibres*

Skeletal muscles are made up of **two types** of **muscle fibres** — **slow twitch** and **fast twitch**. **Different muscles** have **different proportions** of slow and fast twitch fibres. The two types have **different properties**:

SLOW TWITCH MUSCLE FIBRES	FAST TWITCH MUSCLE FIBRES
Muscle fibres that contract slowly.	Muscle fibres that contract very quickly.
Muscles you use for posture, e.g. those in the back, have a high proportion of them.	Muscles you use for fast movement, e.g. those in the eyes and legs, have a high proportion of them.
Good for endurance activities, e.g. maintaining posture, long-distance running.	Good for short bursts of speed and power, e.g. eye movement, sprinting.
Can work for a long time without getting tired.	Get tired very quickly.
Energy's released slowly through aerobic respiration. Lots of mitochondria and blood vessels supply the muscles with oxygen.	Energy's released quickly through anaerobic respiration using glycogen (stored glucose). There are few mitochondria or blood vessels.
Reddish in colour because they're rich in myoglobin — a red-coloured protein that stores oxygen.	Whitish in colour because they don't have much myoglobin (so can't store much oxygen).

Morgan used his good looks and his fast twitch muscle fibres to quickly hail a taxi.

Practice Questions

Q1 Which molecule blocks the actin-myosin binding site in resting muscles?

Q2 What's the name of the bond that's formed when a myosin head binds to an actin filament?

Q3 State three differences between slow and fast twitch skeletal muscle fibres.

Exam Questions

Q1 Rigor mortis is the stiffening of muscles in the body after death. It happens when ATP reserves are exhausted. Explain why a lack of ATP leads to muscles being unable to relax. [3 marks]

Q2 Bepridil is a drug that blocks calcium ion channels. Describe and explain the effect this drug will have on muscle contraction. [3 marks]

What does muscle contraction cost? 80p...

Sorry, that's my favourite sciencey joke so I had to fit it in somewhere — a small distraction before you revisit this page. It's tough stuff but you know the best way to learn it. That's right, shut the book and scribble down what you can remember — if you can't remember much, read it again till you can (and if you can remember loads read it again anyway, just to be sure).

Aerobic Respiration

Energy is needed for muscle contraction — this energy comes from respiration (so you have to learn it, unlucky).

Aerobic Respiration Releases Energy

1) **Aerobic respiration** is the process where a **large amount** of energy is **released** by **splitting glucose** into CO_2 (which is released as a **waste product**) and H_2 (which combines with atmospheric O_2 to produce H_2O).

2) The energy released is **used** to **phosphorylate ADP** to **ATP**. ATP is then used to **provide energy** for all the **biological processes** inside a cell.

3) There are **four stages** in aerobic respiration — **glycolysis**, the **link reaction**, the **Krebs cycle** and **oxidative phosphorylation**.

4) The **first three** stages are a **series of reactions**. The **products** from these reactions are **used** in the **final stage** to produce loads of ATP.

5) **Each reaction** in respiration is **controlled** by a **different enzyme**.

6) The **first** stage happens in the **cytoplasm** of cells and the **other three** stages take place in the **mitochondria**. You might want to refresh your memory of mitochondrion structure before you start.

7) **Coenzymes** are used in respiration, for example:
- **NAD** and **FAD** transfer **hydrogen** from one molecule to another — this means they can **reduce** (give hydrogen to) or **oxidise** (take hydrogen from) a molecule.
- **Coenzyme A** transfers **acetate** between molecules (see pages 55-56).

8) All cells use **glucose** to **respire**, but organisms can also **break down** other **complex organic molecules** (e.g. fatty acids, amino acids), which can then be respired.

Glucose is a respiratory substrate — a molecule that can be respired.

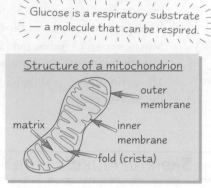

Structure of a mitochondrion

outer membrane
matrix
inner membrane
fold (crista)

See page 5 for what a coenzyme is.

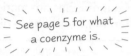

Respiration Map
You are here

Glycolysis
↓
Link Reaction
↓
Krebs Cycle
↓
Oxidative Phosphorylation

Stage 1 — Glycolysis Makes Pyruvate from Glucose

1) Glycolysis involves splitting **one molecule** of **glucose** (a **hexose** sugar, so it has 6 carbons — 6C) into **two** smaller molecules of **pyruvate** (3C).

2) The process happens in the **cytoplasm** of cells.

3) Glycolysis is the **first stage** of both aerobic and anaerobic respiration and **doesn't need oxygen** to take place — so it's an **anaerobic** process.

There are Two Stages in Glycolysis — Phosphorylation and Oxidation

First, **ATP** is **used** to **phosphorylate glucose** to triose phosphate. Then **triose phosphate** is **oxidised**, **releasing ATP**. Overall there's a **net gain** of **2 ATP**.

① Stage One — Phosphorylation

1) Glucose is **phosphorylated** by adding **2 phosphates** from **2 molecules** of **ATP**.

2) This creates **2 molecules** of **triose phosphate** and 2 molecules of ADP.

② Stage Two — Oxidation

1) Triose phosphate is **oxidised** (loses hydrogen), forming **2 molecules** of **pyruvate**.

2) NAD collects the hydrogen ions, forming **2 reduced NAD**.

3) **4 ATP** are **produced**, but 2 were used up in stage one, so there's a **net gain** of **2 ATP**.

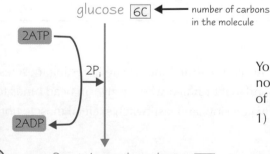

glucose 6C ← number of carbons in the molecule

2ATP
2P$_i$
2ADP

2 × triose phosphate 3C

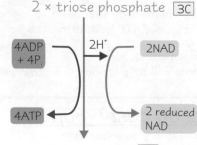

4ADP + 4P$_i$
2H$^+$
2NAD
4ATP
2 reduced NAD

2 × pyruvate 3C

You're probably wondering what now happens to all the products of glycolysis...

1) The **two** molecules of **reduced NAD** go to the **last stage** (oxidative phosphorylation — see page 56).

2) The **two pyruvate** molecules go into the **matrix** of the **mitochondria** for the **link reaction** (see the next page).

Pyruvate is also called pyruvic acid.

Aerobic Respiration

Stage 2 — the Link Reaction converts Pyruvate to Acetyl Coenzyme A

The **link reaction** takes place in the **mitochondrial matrix**:

Respiration Map

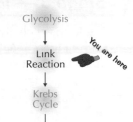

Glycolysis

Link
Reaction

Krebs
Cycle

Oxidative
Phosphorylation

You are here

1) **Pyruvate** is **decarboxylated** (**carbon** is **removed**) — **one carbon atom** is **removed** from pyruvate in the form of CO_2.

2) **NAD** is **reduced** — it collects **hydrogen** from pyruvate, changing **pyruvate** into **acetate**.

3) **Acetate** is combined with **coenzyme A** (CoA) to form **acetyl coenzyme A** (**acetyl CoA**).

4) **No ATP** is produced in this reaction.

pyruvate ☐3C

CO_2 ☐1C

NAD

reduced NAD

acetate ☐2C

coenzyme A
(CoA)

acetyl CoA ☐2C

The Link Reaction occurs Twice for every Glucose Molecule

Two pyruvate molecules are made for **every glucose molecule** that enters glycolysis. This means the **link reaction** and the third stage (the **Krebs cycle**) happen **twice** for every glucose molecule. So for each glucose molecule:

• **Two** molecules of **acetyl coenzyme A** go into the Krebs cycle (see the next page).

• **Two CO_2 molecules** are released as a waste product of respiration.

• **Two** molecules of **reduced NAD** are formed and go to the last stage (oxidative phosphorylation, see the next page).

Practice Questions

Q1 How many stages are there in aerobic respiration?

Q2 Name one coenzyme involved in respiration.

Q3 Where in the cell does glycolysis occur?

Q4 Is glycolysis an anaerobic or aerobic process?

Q5 How many ATP molecules are used up in glycolysis?

Q6 What is the product of the link reaction?

Exam Questions

Q1 During glycolysis a 6-carbon molecule of glucose is converted to pyruvate.
a) Describe the process of glycolysis. [6 marks]
b) What is the overall net gain of ATP from glycolysis? [1 mark]

Q2 The link reaction of respiration occurs in the mitochondrial matrix.
Describe what happens in the link reaction. [3 marks]

No ATP was harmed during this reaction...

Ahhhh... too many reactions. I'm sure your head hurts now, 'cause mine certainly does. Just think of revision as like doing exercise — it can be a pain while you're doing it (and maybe afterwards too), but it's worth it for the well-toned brain you'll have. Just keep going over and over it, until you get the first two stages of respiration straight in your head. Then relax.

Aerobic Respiration

As you've seen, glycolysis produces a net gain of two ATP. Pah, we can do better than that.
The Krebs cycle and oxidative phosphorylation are where it all happens — ATP galore.

Stage 3 — the **Krebs Cycle** Produces **Reduced Coenzymes** and **ATP**

The Krebs cycle involves a series of **oxidation-reduction reactions**, which take place in the **matrix** of the **mitochondria**. Each of these reactions is controlled by a **specific intracellular enzyme**. The cycle happens **once** for **every pyruvate** molecule, so it goes round **twice** for **every glucose** molecule.

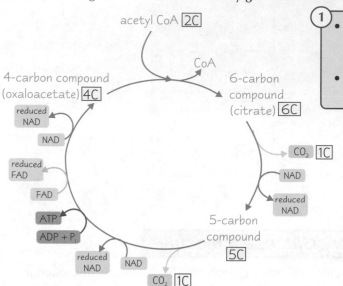

acetyl CoA [2C]

CoA

4-carbon compound (oxaloacetate) [4C]

reduced NAD
NAD

reduced FAD
FAD

ATP
ADP + P$_i$

reduced NAD   NAD

CO$_2$ [1C]

6-carbon compound (citrate) [6C]

CO$_2$ [1C]

NAD

reduced NAD

5-carbon compound [5C]

(1)
- **Acetyl CoA** from the link reaction combines with **oxaloacetate** to form **citrate**.
- **Coenzyme A** goes back to the **link reaction** to be used again.

(2)
- The **6C citrate molecule** is converted to a **5C molecule**.
- **Decarboxylation** occurs, where **CO$_2$** is **removed**.
- **Dehydrogenation** also occurs — this is where **hydrogen** is **removed**.
- The hydrogen is used to **produce reduced NAD** from NAD.

(3)
- The **5C molecule** is then converted to a **4C molecule**. (There are some intermediate compounds formed during this conversion, but you don't need to know about them.)
- **Decarboxylation** and **dehydrogenation** occur, producing **one** molecule of **reduced FAD** and **two** of **reduced NAD**.
- **ATP** is produced by the **direct transfer** of a **phosphate** group from an **intermediate** compound to **ADP**. When a phosphate group is directly transferred from one molecule to another it's called **substrate-level phosphorylation**. Citrate has now been **converted** into **oxaloacetate**.

Respiration Map

Glycolysis

↓

Link Reaction

You are here

Krebs Cycle

↓

Oxidative Phosphorylation

Some **Products** of the **Krebs Cycle** are Used in **Oxidative Phosphorylation**

Some products are **reused**, some are **released** and others are used for the **next stage** of respiration:

Product from one Krebs cycle	Where it goes
1 coenzyme A	Reused in the next link reaction
Oxaloacetate	Regenerated for use in the next Krebs cycle
2 CO$_2$	Released as a waste product
1 ATP	Used for energy
3 reduced NAD	To oxidative phosphorylation
1 reduced FAD	To oxidative phosphorylation

Mr Krebs

Talking about oxidative phosphorylation was always a big hit with the ladies...

Stage 4 — **Oxidative Phosphorylation** Produces *Lots of* **ATP**

1) Oxidative phosphorylation is the process where the **energy** carried by **electrons**, from **reduced coenzymes** (reduced NAD and reduced FAD), is used to **make ATP**. (The whole point of the previous stages is to make reduced NAD and reduced FAD for the final stage.)

2) Oxidative phosphorylation involves two processes — the **electron transport chain** and **chemiosmosis** (see the next page).

Respiration Map

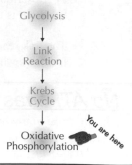
Glycolysis

↓

Link Reaction

↓

Krebs Cycle

↓

Oxidative Phosphorylation
You are here

Aerobic Respiration

Protons are Pumped Across the Inner Mitochondrial Membrane

So now on to how **oxidative phosphorylation** actually **works**:

1) **Hydrogen atoms** are released from **reduced NAD** and **reduced FAD** as they're oxidised to NAD and FAD. The H atoms **split** into **protons (H⁺)** and **electrons (e⁻)**.

2) The **electrons** move along the **electron transport chain** (made up of three **electron carriers**), **losing energy** at each carrier.

3) This energy is used by the electron carriers to **pump protons** from the **mitochondrial matrix into** the **intermembrane space** (the space **between** the inner and outer **mitochondrial membranes**).

4) The **concentration** of **protons** is now **higher** in the **intermembrane space** than in the mitochondrial matrix — this forms an **electrochemical gradient** (a **concentration gradient** of **ions**).

5) Protons **move down** the **electrochemical gradient**, back into the mitochondrial matrix, via **ATP synthase**. This **movement** drives the synthesis of **ATP** from **ADP** and **inorganic phosphate** (P_i).

6) The movement of H⁺ ions across a membrane, which generates ATP, is called **chemiosmosis**.

7) In the mitochondrial matrix, at the end of the transport chain, the **protons**, **electrons** and O_2 (from the blood) combine to form **water**. Oxygen is said to be the final **electron acceptor**.

The regenerated coenzymes are reused in the Krebs cycle.

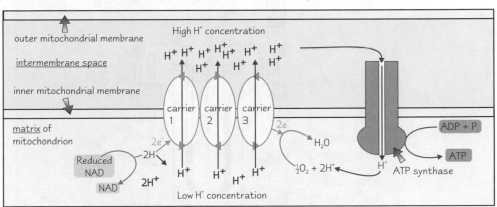

32 ATP Can be Made from One Glucose Molecule

As you know, **oxidative phosphorylation** makes **ATP** using energy from the reduced coenzymes   **2.5 ATP** are made from each **reduced NAD** and **1.5 ATP** are made from each **reduced FAD**. The table on the right shows **how much** ATP a cell can make from **one molecule** of **glucose** in **aerobic respiration**. (Remember, one molecule of glucose produces 2 pyruvate, so the link reaction and Krebs cycle happen twice.)

Stage of respiration	Molecules produced	Number of ATP molecules
Glycolysis	2 ATP	2
Glycolysis	2 reduced NAD	2 × 2.5 = 5
Link Reaction (×2)	2 reduced NAD	2 × 2.5 = 5
Krebs cycle (×2)	2 ATP	2
Krebs cycle (×2)	6 reduced NAD	6 × 2.5 = 15
Krebs cycle (×2)	2 reduced FAD	2 × 1.5 = 3
		Total ATP = 32

The number of ATP produced per reduced NAD or reduced FAD was thought to be 3 and 2, but new research has shown that the figures are nearer 2.5 and 1.5.

Practice Questions

Q1 Where in the cell does the Krebs cycle occur?

Q2 How many carbon dioxide molecules are produced during one turn of the Krebs cycle?

Q3 What do the electrons lose as they move along the electron transport chain in oxidative phosphorylation?

Exam Question

Q1 Carbon monoxide inhibits the final electron carrier in the electron transport chain.
   a) Explain how this affects ATP production via the electron transport chain. [2 marks]
   b) Explain how this affects ATP production via the Krebs cycle. [2 marks]

The electron transport chain isn't just a FAD with the examiners...

Oh my gosh, I didn't think it could get any worse... You may be wondering how to learn these pages of crazy chemistry, but basically you have to put in the time and go over and over it. Don't worry though, it WILL pay off, and before you know it you'll be set for the exam. And once you know this section you'll be able to do anything, e.g. world domination...

Respirometers and Anaerobic Respiration

Congratulations — you've nearly finished respiration. You just need to get to grips with an 'exciting' respiration experiment and the tiny subject of anaerobic respiration, then it's all over. (When I say 'exciting' I'm using the word loosely, but I've got to say something positive to keep the morale up.)

The Rate of Respiration can be Measured using a Respirometer

1) The volume of **oxygen taken up** or the volume of **carbon dioxide produced indicates** the **rate** of **respiration**.

2) A **respirometer** measures the rate of **oxygen** being **taken up** — the **more** oxygen taken up, the **faster** the rate of respiration.

3) Here's how you can use a **respirometer** to **measure** the volume of **oxygen taken up** by some **woodlice**:

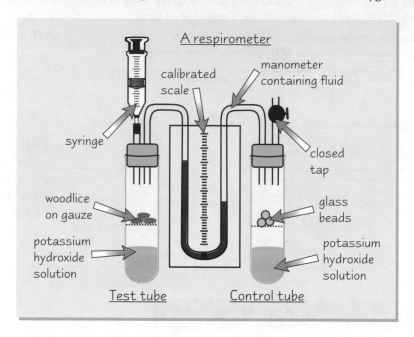

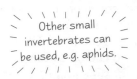
Other small invertebrates can be used, e.g. aphids.

Alfred the aphid thought holding his breath in the respirometer would be really funny. The students didn't.

- The apparatus is set up as shown above.

- **Each tube** contains **potassium hydroxide** solution (or soda lime), which **absorbs carbon dioxide**.

- The **control tube** is set up in exactly the **same way** as the test tube, but **without** the **woodlice**, to make sure the **results** are **only** due to the woodlice **respiring** (e.g. it contains beads that have the same mass as the woodlice).

- The **syringe** is used to set the **fluid** in the **manometer** to a **known level**.

- The apparatus is **left** for a **set** period of **time** (e.g. 20 minutes).

- During that time there'll be a **decrease** in the **volume** of the **air** in the test tube, due to **oxygen consumption** by the **woodlice** (all the CO_2 produced is absorbed by the potassium hydroxide).

- The decrease in the volume of the air will **reduce the pressure** in the tube and cause the **coloured liquid** in the manometer to **move towards** the test tube.

- The **distance moved** by the **liquid** in a **given time** is **measured**. This value can then be used to **calculate** the **volume of oxygen** taken in by the woodlice **per minute**.

- Any **variables** that could **affect** the results are **controlled**, e.g. temperature, volume of potassium hydroxide solution in each test tube.

4) To produce more **reliable** results the experiment is **repeated** and a **mean volume** of O_2 is calculated.

Respirometers and Anaerobic Respiration

Lactate Fermentation is a Type of Anaerobic Respiration

1) **Anaerobic** respiration **doesn't use oxygen**.

2) It **doesn't** involve the **link reaction**, the **Krebs cycle** or **oxidative phosphorylation**.

3) There are **two types** of anaerobic respiration, but you only need to know about one of them — **lactate fermentation**.

4) Lactate fermentation occurs in **animals** and produces **lactate**.

5) Here's how it works:

Some bacteria carry out lactate fermentation.

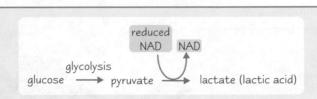

glucose → pyruvate → lactate (lactic acid)
glycolysis
reduced NAD   NAD

- **Glucose** is converted to **pyruvate** via **glycolysis**.
- **Reduced NAD** (from glycolysis) transfers **hydrogen** to **pyruvate** to form **lactate** and **NAD**.
- **NAD** can then be reused in **glycolysis**.

6) The production of lactate **regenerates NAD**. This means **glycolysis** can **continue** even when there **isn't** much oxygen around, so a **small amount of ATP** can still be **produced** to keep some biological process going... clever.

Lactic Acid Needs to be Broken Down

After a period of anaerobic respiration **lactic acid builds up**. Animals can **break down lactic acid** in **two** ways:

1) **Cells** can **convert** the lactic acid back to **pyruvate** (which then re-enters aerobic respiration at the **Krebs cycle**).

2) **Liver cells** can **convert** the lactic acid back to **glucose** (which can then be **respired** or **stored**).

Practice Questions

Q1 What does a respirometer measure?

Q2 Lactate fermentation is an example of what type of respiration?

Q3 Give one way that animals can break down lactate.

Exam Questions

Q1 A respirometer is set up as shown in the diagram on the previous page.
a) Explain the purpose of the control tube. [1 mark]
b) Explain what would happen if there was no potassium hydroxide in the tubes. [2 marks]
c) What other substance could be measured to find out the rate of respiration? [1 mark]

Q2 A culture of mammalian cells was incubated with glucose, pyruvate and antimycin C.
Antimycin C inhibits an electron carrier in the electron transport chain of aerobic respiration.
Explain why these cells can still produce lactate. [1 mark]

Lactic acid — visibly present after extreme exertion on a bouncy castle...

Okay, that wasn't very funny, but these pages don't really give me any inspiration. You probably feel the same way. It's just one of those times where you have to plough through them. You could try drawing a few pretty diagrams to get the experiment or fermentation reaction in your head. Then you could do something exciting, like sticking your toe in your ear...

Electrical Activity in the Heart

Your heart rate increases with exercise. Doctors can monitor the heart with a fancy machine to check if it's beating OK and to check for signs of disease. Beep, beep, beep, beep, beep, beeeeeeeeeeeep.... uh oh...

Cardiac Muscle Controls the Regular Beating of the Heart

Cardiac (heart) muscle is '**myogenic**' — it can contract and relax without receiving signals from neurones. This pattern of contractions controls the **regular heartbeat**.

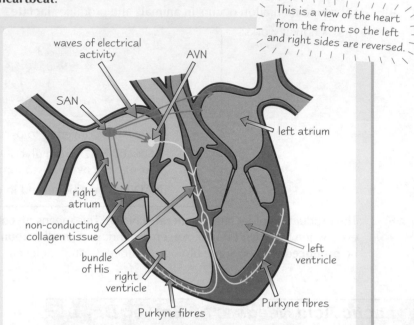

This is a view of the heart from the front so the left and right sides are reversed.

1) The process starts in the **sinoatrial node (SAN)**, which is in the wall of the **right atrium**.

2) The SAN is like a pacemaker — it sets the **rhythm** of the heartbeat by sending out regular **waves of electrical activity** to the **atrial walls**.

3) This causes the right and left **atria** to **contract at the same time**.

4) A band of non-conducting **collagen tissue** prevents the waves of electrical activity from being passed directly from the atria to the ventricles.

5) Instead, these waves of electrical activity are transferred from the SAN to the **atrioventricular node (AVN)**.

6) The AVN is responsible for passing the waves of electrical activity onto the bundle of His. But, there's a **slight delay** before the AVN reacts, to make sure the ventricles contract **after** the atria have emptied.

7) The **bundle of His** is a group of muscle fibres responsible for conducting the waves of electrical activity to the finer muscle fibres in the right and left **ventricle walls**, called the **Purkyne fibres**.

8) The Purkyne fibres carry the waves of electrical activity into the muscular walls of the right and left ventricles, causing them to **contract simultaneously**, from the bottom up.

An Electrocardiograph Records the Electrical Activity of the Heart

A doctor can check someone's **heart function** using an **electrocardiograph** — a machine that **records** the **electrical activity** of the heart. The heart muscle **depolarises** (loses electrical charge) when it **contracts**, and **repolarises** (regains charge) when it **relaxes**. An electrocardiograph records changes in electrical charge using **electrodes** placed on the chest.

The trace produced by an electrocardiograph is called an **electrocardiogram**, or **ECG**. A **normal** ECG looks like this:

1) The **P wave** is caused by **contraction** (depolarisation) of the **atria**.

2) The main peak of the heartbeat, together with the dips at either side, is called the **QRS complex** — it's caused by **contraction** (depolarisation) of the **ventricles**.

3) The **T wave** is due to **relaxation** (repolarisation) of the **ventricles**.

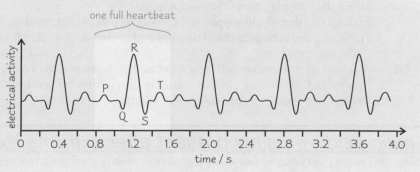

Electrical Activity in the Heart

Doctors use ECGs to Diagnose Heart Problems

Doctors **compare** their patients' ECGs with a **normal trace**. This helps them to **diagnose** any **problems** with the heart's **rhythm**, which may indicate **cardiovascular disease** (heart and circulatory disease). Here are some examples of abnormal traces:

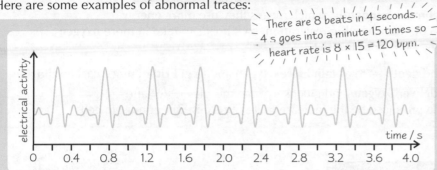

There are 8 beats in 4 seconds. 4 s goes into a minute 15 times so heart rate is 8 × 15 = 120 bpm.

Tachycardia — increased heart rate

Here, the heartbeat is **too fast**, around **120** beats per minute. This could be a sign of **heart failure** — a problem with the heart means that it **can't pump blood efficiently**, so heart rate **increases** to ensure **enough blood** is pumped around the body. Tachycardia can also increase the **risk of a heart attack**.

Problem with the AVN

Here, the **atria** are contracting but the **ventricles** are **not** (some **P** waves aren't followed by a **QRS** complex). This might mean there's a problem with the **AVN** — impulses aren't travelling from the atria through to the ventricles.

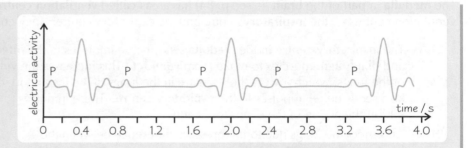

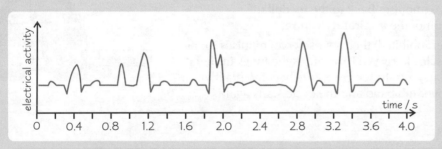

Fibrillation — irregular heart beat

The heart beat in this ECG is **irregular**. Both the **atria** or **ventricles** have lost their rhythm and **stopped contracting properly**. Atrial fibrillation can lead to **chest pains**, **fainting** and an **increased risk** of stroke. Ventricular fibrillation can quickly **cause death**. It may be **caused** by a **heart attack**.

Practice Questions

Q1 What prevents impulses from the atria travelling straight into the ventricles?

Q2 What is the name of the structure that picks up impulses from the atria and passes them on to the ventricles?

Q3 What causes the QRS part of the ECG trace?

Exam Questions

Q1 Describe the function of:

   a) the sinoatrial node. [1 mark]

   b) the bundle of His. [1 mark]

Q2 Suggest the cause of an ECG which has a QRS complex that is smaller than normal. [2 marks]

Perhaps if I plug myself into the mains, my heart'll be supercharged...

It's pretty incredible that your heart manages to go through all those stages in the right order, at exactly the right time, without getting it even slightly wrong. It does it perfectly, about 70 times every minute. That's about 100 800 times a day. If only my brain was that efficient. I'd have all this revision done in five minutes, then I could go and watch TV...

Variations in Heart Rate and Breathing Rate

Your heart doesn't beat away steadily all day — heart rate increases and decreases, depending on how active you are.
Whether you're dozing on the sofa or going on a bike ride, how active you are also affects your breathing rate...

Breathing Rate and Heart Rate Increase When you Exercise

When you exercise your **muscles contract more frequently**, which means they use **more energy**.
To replace this energy your body needs to do **more aerobic respiration**, so it needs to **take in more oxygen**
and **breathe out more carbon dioxide**. The body does this by:

1) **Increasing breathing rate** and **depth** — to **obtain more oxygen** and to **get rid** of **more carbon dioxide**.

2) **Increasing heart rate** — to **deliver oxygen** (and glucose) to the muscles **faster** and **remove extra carbon dioxide** produced by the increased rate of **respiration** in muscle cells.

The ventilation centre is also called the respiratory centre.

The Medulla Controls Breathing Rate

The **medulla** (a part of the **brain** — see p. 84) has areas called **ventilation centres**. There are two
ventilation centres — the **inspiratory** centre and the **expiratory** centre. They control the **rate of breathing**:

1) The **inspiratory centre** in the **medulla** sends nerve impulses to the **intercostal**
and **diaphragm** muscles to make them **contract**. This **increases** the **volume**
of the lungs, which **lowers** the **pressure** in the lungs. (The inspiratory centre
also sends nerve impulses to the **expiratory centre**. These impulses
inhibit the action of the **expiratory centre**.)

2) **Air enters** the lungs due to the **pressure difference** between the
lungs and the air outside.

3) As the **lungs inflate**, **stretch receptors** in the lungs are **stimulated**.
The stretch receptors send nerve impulses back to the **medulla**.
These impulses **inhibit** the action of the **inspiratory centre**.

4) The expiratory centre (no longer inhibited) then sends nerve impulses to the
diaphragm and **intercostal muscles** to **relax**. This causes the **lungs to deflate**,
expelling air. As the lungs deflate, the **stretch receptors** become **inactive**.
The inspiratory centre is no longer inhibited and the cycle starts again.

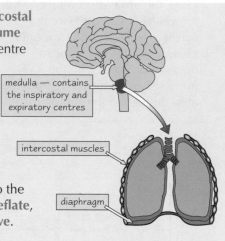

medulla — contains
the inspiratory and
expiratory centres

intercostal muscles

diaphragm

Exercise Triggers an Increase in Breathing Rate by Decreasing Blood pH

1) During exercise, the level of **carbon dioxide** (CO_2) in the blood **increases**.
This **decreases** the **pH** of the blood.

2) There are **chemoreceptors** (receptors that sense chemicals) in the **medulla**, **aortic bodies**
(in the aorta) and **carotid bodies** (in the carotid arteries carrying blood to the brain)
that are **sensitive** to **changes in blood pH**.

3) If the chemoreceptors **detect** a **decrease** in blood **pH**, they send nerve impulses to
the **medulla**, which sends **more frequent** nerve impulses to the **intercostal muscles**
and **diaphragm**. This **increases** the **rate** and **depth** of breathing.

4) This causes **gaseous exchange** to **speed up** — the CO_2 level drops and extra O_2 is
supplied for the muscles.

Ventilation Rate Increases with Exercise

1) Ventilation rate is the **volume** of air **breathed in or out** in a **period of time**, e.g. a minute.

2) It increases during exercise because **breathing rate** and **depth increase**.

Variations in Heart Rate and Breathing Rate

The Medulla Controls Heart Rate Too

Heart rate is controlled by the cardiovascular control centre in the medulla of the brain:

Decreased blood pH causes an increase in heart rate

1) A decrease in blood pH (caused by an increase in CO_2) is detected by chemoreceptors.
2) The chemoreceptors send nerve impulses to the medulla.
3) The medulla sends nerve impulses to the SAN to increase the heart rate.

Increased blood pressure causes a decrease in heart rate

1) Pressure receptors in the aorta wall and in the carotid sinuses (at the start of the carotid arteries carrying blood to the brain) detect changes in blood pressure.
2) If the pressure is too high, the pressure receptors send nerve impulses to the cardiovascular centre, which sends nerve impulses to the SAN, to slow down the heart rate.
3) If the pressure is too low, pressure receptors send nerve impulses to the cardiovascular centre, which sends nerve impulses to, yep you guessed it, the SAN, to speed up the heart rate.

Derek wasn't sure if his heart rate had increased because of running or the fact that Janice was wearing a thong.

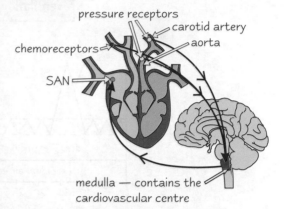

pressure receptors
carotid artery
chemoreceptors
aorta
SAN
medulla — contains the cardiovascular centre

Exercise Triggers an Increase in Heart Rate by Decreasing Blood pH

During exercise, the level of carbon dioxide (CO_2) in the blood increases. This decreases the pH of the blood, which the chemoreceptors detect. The leads to an increase in heart rate (see above).

Cardiac Output Increases with Exercise

1) Cardiac output is the total volume of blood pumped by a ventricle every minute.
2) The equation for working out cardiac output is:

> Cardiac output (cm^3/min) = heart rate (beats per minute) × stroke volume (cm^3)

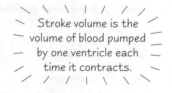

Stroke volume is the volume of blood pumped by one ventricle each time it contracts.

3) So cardiac output increases during exercise because heart rate increases (stroke volume also increases because the heart pumps harder as well).

Practice Questions

Q1 Which part of the brain controls breathing rate and heart rate?

Q2 What effect does exercise have on cardiac output?

Exam Question

Q1 In a laboratory experiment, an animal was anaesthetised and dilute carbonic acid (carbon dioxide in solution) was added to the blood in the coronary artery.

a) What effect would you expect this to have on the animal's breathing rate? Explain your answer. [5 marks]

b) Cardiac output increased in response to the addition of carbonic acid. How is cardiac output calculated? [1 mark]

My heart rate increases when Ronaldo exercises...

...because I find him incredibly annoying, not because he's attractive in any way. Breathing rate and heart rate can be increased to supply more oxygen for aerobic respiration, which releases the energy the body needs during exercise. Cardiac output increases as heart rate (and stroke volume) increases, so a greater volume of blood is pumped around the body.

Investigating Ventilation

You need to know how to investigate the effects of exercise on all things to do with ventilation. And I'm not talking about the ventilation that Bruce Willis is fond of climbing through in films...

Tidal Volume is the Volume of Air in a Normal Breath

Here are some terms that you need to know about breathing:

1) **Tidal volume** — the **volume** of air in **each breath**, usually about **0.4 dm³**.

2) **Breathing rate** — **how many breaths** are taken, usually in a **minute**.

3) **Ventilation rate** — the **volume** of air **breathed in or out**, usually in a **minute**.
Here's how it's calculated:

(see page 62)

> **ventilation rate = tidal volume × breathing rate**

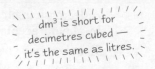

dm³ is short for decimetres cubed — it's the same as litres.

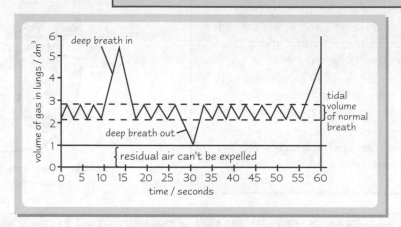

Jane couldn't maintain her breathing rate when she saw all those TVs.

Spirometers Can be Used to Measure Tidal Volume and Breathing rate

A spirometer is a machine that can give readings of **tidal volume** and **breathing rate**.

1) A spirometer has an **oxygen-filled** chamber with a **movable lid**.

2) A person breathes through a **tube** connected to the oxygen chamber.

3) As the person breathes **in** the lid of the chamber moves **down**. When they breathe **out** it moves **up**.

4) These movements are recorded by a **pen** attached to the lid of the chamber — this writes on a **rotating drum**, creating a **spirometer trace**.

5) The **soda lime** in the tube the person breathes into absorbs **carbon dioxide**.

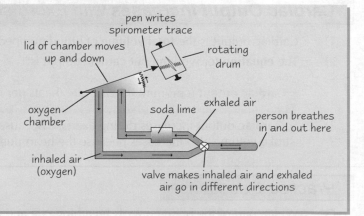

The **total volume of gas** in the chamber **decreases** over time. This is because the air that's breathed out is a **mixture** of oxygen and carbon dioxide. The carbon dioxide is absorbed by the **soda lime** — so there's **only oxygen** in the chamber which the person inhales from. As this oxygen gets used up by respiration, the total volume decreases.

Spirometers Can be Used to Investigate the Effects of Exercise

Exercise causes an increase in **breathing rate** (see page 62) and **tidal volume**. You can use a spirometer to measure the change in breathing rate and tidal volume at **rest**, **during exercise** and **after exercise**. For example:

1) A person is connected to a spirometer using a **mask** so that continuous readings can be recorded.

2) Readings are recorded for **one minute** at **rest**.

3) The person then begins to **exercise**, e.g. running on a **treadmill**, for **two minutes**.

4) The person then **stops exercising** and readings are continued for **one minute** at **rest**.

Investigating Ventilation

You Need to be Able to *Analyse Data* from a *Spirometer*

You can use spirometer traces to look at the **effect** of **exercise** on breathing rate and tidal volume. Here's an example:

EXAMPLE 1

1) A person's **breathing rate** and **tidal volume** were measured using a **spirometer** and the method described on the previous page. The spirometer trace is shown in the **graph** below.

2) At **rest**, tidal volume is about **0.4 dm³** and breathing rate is **12 breaths per minute**.

3) During **exercise**, the body needs **more oxygen** for muscle contraction and it needs to **remove more carbon dioxide** (see page 62). So breathing rate and tidal volume **increase** — to around **20 breaths per minute** and **3.2 dm³**.

4) During **recovery**, the body still needs to keep **breathing hard** (to get oxygen to remove any lactate that's built up), but eventually breathing rate and tidal volume return to **rest levels**.

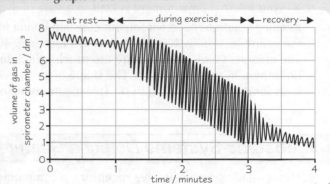

This graph looks different to the one on the previous page because it shows the volume of air in the spirometer, not in the lungs.

You can also look at the **effect** of a **fitness training programme** on breathing rate and tidal volume:

EXAMPLE 2

A person's **breathing rate** and **tidal volume** were measured using a **spirometer** and the method described on the previous page. The results are shown in the **graph** below. The **same test** was repeated three months later on the **same person**, after they'd gone through a **fitness training programme**. The spirometer results were plotted to **compare** the **effects** of training:

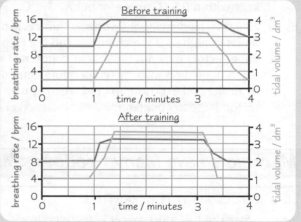

1) Training **decreased breathing rate at all stages** because the lung muscles are **strengthened**, so **more air** is taken in with each breath, meaning **fewer breaths** are needed.

2) Training **increased tidal volume at all stages**, again because muscles are strengthened, so more air is taken in with each breath.

3) During **recovery**, **breathing rate** and **tidal volume** **decreased faster** due to training because the muscles are strengthened, so the lungs can get oxygen and carbon dioxide supplies back to normal quicker.

Practice Questions

Q1 What is tidal volume?

Q2 What's the purpose of soda lime in a spirometer?

Exam Question

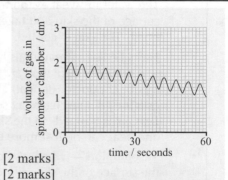

Q1 The graph on the right shows a spirometer trace taken from a student at rest.

a) Calculate the tidal volume and breathing rate. [2 marks]
b) How would you expect the trace to differ if the student was exercising? [2 marks]

Investigate someone's breathing — make sure they've had a mint first...

I thought spirometers were those circular plastic things you draw crazy patterns with... apparently not. I know the graphs don't look that approachable, but it's important you understand what the squiggly lines show, and the meaning of terms used when investigating breathing — I'd bet my right lung there'll be a question on spirometer graphs in the exam.

Homeostasis

Whilst you're pounding on the treadmill, behind the scenes your body is working hard to control your body temperature.

Homeostasis is the Maintenance of a Constant Internal Environment

1) Your **external environment** and **what you're doing** (e.g. exercising) can affect your **internal environment** — the blood and tissue fluid that surrounds your cells. For example, **exercise increases body temperature**.

2) **Homeostasis** involves **control systems** that keep your **internal environment** roughly **constant** — your internal environment is kept in a state of **dynamic equilibrium** (i.e. fluctuating around a normal level).

3) **Keeping** your internal environment **constant** is vital for cells to **function normally** and to **stop** them being **damaged**. For example, if **body temperature** is **too high** (e.g. 40 °C) **enzymes** may become **denatured**. The enzyme's molecules **vibrate too much**, which **breaks** the **hydrogen bonds** that hold them in their **3D shape**. The **shape** of the enzyme's **active site** is **changed** and it **no longer works** as a **catalyst**. This means **metabolic reactions** are **less efficient**.

Things like blood pH and glucose concentration are also maintained by homeostasis.

Homeostatic Systems Detect a Change and Respond by Negative Feedback

1) Homeostatic systems involve **receptors**, a **communication system** and **effectors** (see p. 72).

2) Receptors detect when a level is **too high** or **too low**, and the information's communicated via the **nervous** system or the **hormonal** system to **effectors**.

3) The effectors respond to **counteract** the change — bringing the level **back** to **normal**.

4) The mechanism that **restores** the level to **normal** is called a **negative feedback** mechanism.

5) Negative feedback **keeps** things around the **normal** level, e.g. body temperature is usually kept **within 0.5 °C** above or below **37 °C**.

6) Negative feedback only works within **certain limits** though — if the change is **too big** then the **effectors** may **not** be able to **counteract** it, e.g. a huge drop in body temperature caused by prolonged exposure to cold weather may be too large to counteract.

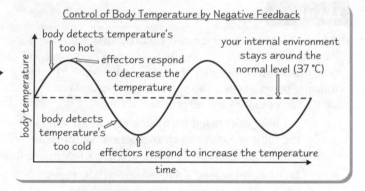

Control of Body Temperature by Negative Feedback

Mammals have Many Feedback Mechanisms to Change Body Temperature

TO REDUCE BODY TEMPERATURE:

- <u>Sweating</u> — **more sweat** is secreted from **sweat glands** when the body's too hot. The water in sweat **evaporates** from the surface of the skin and **takes heat** from the body. The **skin is cooled**.

- <u>Hairs lie flat</u> — mammals have a layer of **hair** that provides **insulation** by **trapping air** (air is a poor conductor of heat). When it's hot, **erector pili muscles relax** so the hairs lie flat. **Less air** is trapped, so the skin is **less insulated** and **heat** can be **lost** more easily.

- <u>Vasodilation</u> — when it's hot, **arterioles** near the surface of the skin **dilate** (this is called **vasodilation**). **More blood** flows through the **capillaries** in the surface layers of the dermis. This means **more heat is lost** from the skin by **radiation** and the **temperature is lowered**.

TO INCREASE BODY TEMPERATURE:

- <u>Shivering</u> — when it's cold, **muscles contract** in **spasms**. This makes the body **shiver** and **more heat is produced** from **increased respiration**.

- <u>Hormones released</u> — the body releases **adrenaline** and **thyroxine**, which **increase metabolism** so **more heat is produced**.

- <u>Much less sweat</u> — less sweat is secreted from sweat glands when it's cold, **reducing the amount of heat loss**.

- <u>Hairs stand up</u> — **erector pili muscles contract** when it's cold, which makes the **hairs stand up**. This **traps more air** and so **prevents heat loss**.

- <u>Vasoconstriction</u> — when it's cold, **arterioles** near the surface of the skin **constrict** (this is called **vasoconstriction**) so **less blood** flows through the **capillaries** in the surface layers of the dermis. This **reduces heat loss**.

Homeostasis

The **Hypothalamus Controls** Body Temperature in **Mammals**

1) **Body temperature** in mammals is **maintained** at a **constant level** by a part of the **brain** called the **hypothalamus**.

2) The hypothalamus **receives information** about **temperature** from **thermoreceptors** (temperature receptors).

3) Thermoreceptors send **impulses** along **sensory neurones** to the **hypothalamus**, which sends **impulses** along **motor neurones** to **effectors** (muscles and glands).

4) The effectors respond to **restore** the body temperature **back to normal**.

5) The control of body temperature is called **thermoregulation**. Here's how it all works:

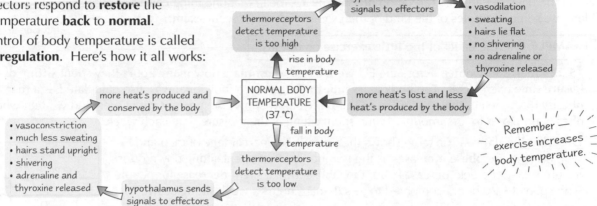

Remember — exercise increases body temperature.

Hormones Switch Genes On to **Help Regulate Temperature**

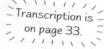

Transcription is on page 33.

1) In a cell there are **proteins** called **transcription factors** that **control** the **transcription** of **genes**.

2) Transcription factors **bind** to **DNA sites** near the **start** of genes and **increase** or **decrease** the **rate** of transcription. Factors that **increase** the rate are called **activators** and those that **decrease** the rate are called **repressors**.

3) **Hormones** can **bind** to some **transcription factors** to **change body temperature**. Here's how:

- At **normal** body temperature, the **thyroid hormone receptor** (a transcription factor) binds to DNA at the **start** of a gene.

- This **decreases** the **transcription** of a gene coding for a **protein** that increases **metabolic rate**.

- At **cold** temperatures **thyroxine** is released, which **binds** to the thyroid hormone receptor, causing it to act as an **activator**.

- The **transcription rate increases**, producing **more protein**. The protein **increases** the **metabolic rate**, causing an increase in **body temperature**.

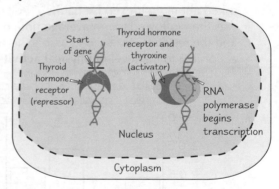

Practice Questions

Q1 What is homeostasis and why is it necessary?

Q2 How does vasoconstriction help to increase body temperature?

Q3 Which part of the brain is responsible for maintaining a constant body temperature in mammals?

Q4 What does a transcription factor do?

Exam Questions

Q1 Describe the role of receptors, communication systems and effectors in homeostasis. [3 marks]

Q2 a) Describe how homeostasis reduces body temperature during exercise. [3 marks]
b) Give two mechanisms that can reduce body temperature. [2 marks]

Homeostasis works like a teacher — everything always gets corrected...

The key to understanding homeostasis is to get your head around negative feedback. Basically, if one thing goes up, the body responds to bring it down — and vice versa. So sweat might be pongy, but you'd be a bit too hot without it.

Exercise and Health

I hope you didn't think you'd get through a lovely section like this without some data analysis to spice things up a bit.

Not Doing Enough Exercise can be Unhealthy...

There are loads of **studies** out there that have looked at the **effects** of **not doing enough exercise**. You might have to analyse some **data** from a study in your exam — so you need to be able to **describe** what the data shows and say if there's a **correlation** (a relationship between two variables, see p. 96). You need to be careful what you **conclude** about the data, because a correlation **doesn't** always mean that one thing **causes** another.
Here are some **examples** of the kind of things you might get in your **exam**:

See pages 96-98 for more on analysing data.

EXAMPLE 1 — The effect of **too little exercise** on **obesity**

15 239 men and **women** across the EU were asked to **estimate** how many **hours** they spent **sitting during their leisure time** each week. The **body mass index** (**BMI**) of each individual was also calculated as a measure of **obesity** (BMI greater than 30 kg / m²). The **table** below shows the **percentage** of men and women who are **obese compared** to the amount of time spent sitting in their leisure time each week.

<u>Describe the data</u> — The table shows that **overall**, the **percentage** of men and women who are **obese increases** as the number of hours spent sitting down during leisure time per week **increases**. But the table shows a **slight decrease** for people who spent **15-20 hours** compared to less than 15 hours a week sitting down.

<u>Draw conclusions</u> — The table shows there's a **correlation** (link) between sitting down for a long time each week and being **obese**, for both men and women. But you **can't** say that long periods of sitting **causes** obesity. There could be **other reasons** for the trend, e.g. people may sit down more **because** they're obese.

Hours sitting down per week	% obese men	% obese women
<15	7.6	9.2
15-20	7.3	6.5
21-25	8.3	10.3
26-35	9.0	11.8
>35	13.3	12.4

EXAMPLE 2 — The effect of **too little exercise** on **coronary heart disease (CHD)**

5159 men aged **40 to 59** with **no history of CHD** were asked about their **exercise habits**. The health of the men was then followed for an **average time** of **16.8 years**, and the results were used to assess the **risk of CHD** according to the level of **physical activity** carried out.

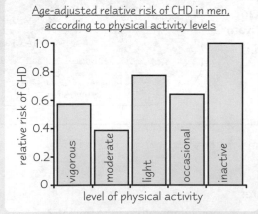

Age-adjusted relative risk of CHD in men, according to physical activity levels

<u>Describe the data</u> — Men aged 40 to 59 who are **inactive** have a **higher risk** of CHD than men of the same age who are physically active. **Moderate exercise** gives the **lowest relative risk** of CHD, at about 0.39.

<u>Draw conclusions</u> — There's a **weak correlation** between **too little physical activity** and an **increased risk** of CHD in **men** aged 40 to 59, but it **doesn't** steadily increase with the amount of exercise. The study **only** involved **men**, so you **can't** say that the same risk occurs in **women**.

EXAMPLE 3 — The effect of **too little exercise** on **Type 2 diabetes**

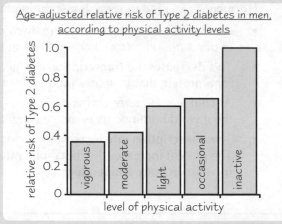

Age-adjusted relative risk of Type 2 diabetes in men, according to physical activity levels

5159 men aged **40 to 59** with **no history of Type 2 diabetes** were asked about their **exercise habits**. The health of the men was then followed for an **average time** of **16.8 years**, and the results were used to assess the **risk of developing Type 2 diabetes** according to the level of **physical activity** carried out.

<u>Describe the data</u> — The **relative risk** of Type 2 diabetes **increases** with **more inactivity** in men aged 40 to 59.

<u>Draw conclusions</u> — There's a **correlation** between **too little physical activity** and an **increased risk** of Type 2 diabetes in **men** aged 40-59. But, you can't say one **causes** the other as there could be **other reasons**, e.g. diet.

Exercise and Health

... But Doing *Too Much Exercise* can be *Unhealthy* Too

It's thought that **too much exercise** could possibly cause some problems too, e.g. wear and tear of the joints. So you might get asked to interpret **data** looking at the **effects** of **too much exercise** as well...

EXAMPLE 1 — The effect of **too much exercise** on **wear** and **tear** of **joints**

The number of **hospital admissions** for **osteoarthritis** of the **hip, knee** and **ankle joints** for **2049** former **male elite athletes** and **1403 healthy, fit controls** were compared by analysing **hospital records** from 1970 to 1990.

<u>Describe the data</u> — The percentage of **male former elite athletes** admitted to hospital for **osteoarthritis** (wear and tear of the joints) is **more than twice** that of healthy men (5.9% compared to 2.6%). The table shows similar high percentages for different kinds of athletes, with **endurance athletes** being **worse affected** (6.8% admitted to hospital).

<u>Draw conclusions</u> — The table shows a **correlation** between being an **elite male athlete** of any kind and having **osteoarthritis** of the hip, knee or ankle. But you can't say that doing a lot of exercise **causes** osteoarthritis — there may be **other reasons** for the trend, e.g. elite athletes may be more likely to injure themselves in competitions, which could lead to arthritis.

Group of men	% of men admitted to hospital for osteoarthritis of the hip, knee or ankle
Healthy men, fully fit for military service	2.6
Former elite athletes (all)	5.9
• Endurance athletes (e.g. long-distance runners)	6.8
• Mixed sports athletes (e.g. footballers)	5.0
• Power sports athletes (e.g. boxers)	6.6

EXAMPLE 2 — The effect of **too much exercise** on the **immune system**

The number of people showing **two or more** symptoms of **upper respiratory tract infection** was recorded for a group containing **32 elite** and **31 recreational triathletes** and **cyclists** during a **five-month** period in their training season. It was also recorded for **20 control** subjects during a five-month period.

<u>Describe the data</u> — There was a **much higher** number of cases of respiratory illnesses in **elite athletes** (21) than in athletes (7) and sedentary controls (9) over the five-month period.

<u>Draw conclusions</u> — There's a **correlation** between doing **a lot of exercise** (being an elite athlete) and getting **more** cases of **respiratory illnesses**. But there's also a **correlation** between doing **some exercise** (recreationally competitive athletes) and getting **fewer** cases of respiratory illnesses. You **can't** say doing a lot of exercise **causes** more respiratory illnesses — there could be **other reasons** for the trend, e.g. elite athletes might be exposed to lots of infections when competing a lot.

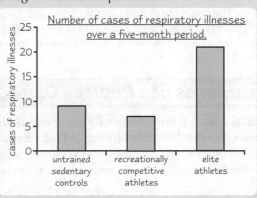

Practice Questions

Q1   What is a correlation?

Exam Question

Physical activity (hours/week)	≥3.5	1–3.5	<1
Relative risk of CHD	1.0	1.32	1.48

Q1   A study investigated the effects of physical activity on the risk of coronary heart disease (CHD) in 88 393 women.
      The results are shown in the table.
   a) Describe the results.   [1 mark]
   b) From the table, what can you conclude about the effects of physical activity on CHD?   [1 mark]

Drawing conclusions — you'll need wax crayons and some paper...

These pages give you some examples to help you deal with what the examiners are sure to hurl at you — they really love throwing data around. There's some important advice here (even if I say so myself) — it's easy to leap to a conclusion that isn't really justified. So make sure you know that just because things are correlated, it doesn't mean one causes the other.

Exercise and Health

Yoga on an exotic beach somewhere, that's my kind of exercise. Apparently not all sports are as calm or as stress-free. There are injuries galore...

Surgery can Help People with Injuries to Play Sports

Some injuries can cause **permanent damage** to the body, e.g. head or spinal injuries. But people can **recover** from some injuries if they're **treated correctly**. A lot of injuries happen when **playing sports** because the body's put under a lot of **stress** (fast running, hard tackles, etc.). Advances in **medical technology** can help people with an injury to **recover** and **participate in sports**. One advance you need to know about is **keyhole surgery**:

1) Keyhole surgery is a way of doing surgery **without** making a **large incision** (cut) in the skin.

2) Surgeons make a much **smaller incision** in the patient, and they insert a tiny **video camera** and **specialised medical instruments** through the incision into the body.

3) There are many **advantages** of keyhole surgery over regular surgery:

- Operations don't involve opening up the patient as much, so patients **lose less blood** and have **less scarring** of the skin.
- Patients are usually in **less pain** after their operation and they **recover more quickly**, because less damage is done to the body.
- This makes it **easier** for the patient to return to **normal activities** and their **hospital stay** is **shorter**.

Debra knew she'd have to do some serious keyhole surgery to see what was in the case.

For example, damaged **cruciate ligaments** can be fixed by **keyhole surgery**:

1) A **common sports injury** is damage to the cruciate ligaments — **ligaments** found in the middle of your **knee**, connecting your **thigh bone** to your **lower leg bone**.

2) **Damaged** cruciate ligament can be **removed** and **replaced** with a **graft** of ligament through a small incision in the knee.

Injuries aren't usually fixed by surgery alone — other treatments (e.g. physiotherapy, anti-inflammatory drugs) are needed too for a full recovery.

Prostheses can Replace Damaged Body Parts

Some people are **born without** a particular **body part**, e.g. without a leg. Other people suffer **injuries** that result in them **losing** or **badly damaging** a body part, e.g. tennis players can damage their knees so much that they can no longer play sports. Sometimes it's possible to **replace** damaged or missing body parts with an **artificial device** called a **prosthetic**:

1) Prostheses can be used to **replace whole limbs** (e.g. an artificial leg can replace a missing leg) or **parts of limbs** (e.g. artificial hip joints can replace damaged hip joints).

2) Some prostheses include **electronic devices** that **operate** the prosthesis by picking up information sent by the **nervous system** (e.g. artificial hand prostheses with an electronic device allow the user to move the fingers).

3) So prostheses make it possible for people with some **disabilities** to **participate** in **sport**, e.g. prosthetic 'legs' (called blades) allow people without legs to run.

4) They also make it possible for people who have certain **injuries** to **play sport again**.

For example, damaged **knee joints** can be replaced by **prosthetic joints**:

1) A **metal device** is inserted into the knee to **replace damaged cartilage** and **bone**.

2) The knee joint and the ends of the leg bones are replaced to provide a **smooth knee joint**. **Cushioning** in the new joint helps to **reduce** the **impact** on the knee.

3) A knee joint replacement allows people with serious knee problems to **move around** and participate in **low-impact sports**, such as walking and swimming.

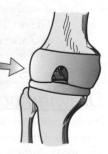

Exercise and Health

Some Athletes Use *Performance-Enhancing Drugs*

When involved in a very **competitive sport**, some people choose to take **performance-enhancing drugs** — these are drugs that will **improve** a person's **performance**. There are various kinds of performance-enhancing drugs that have different effects on the body, for example:

- **Anabolic steroids** — these drugs **increase strength**, **speed** and **stamina** by increasing **muscle size** and allowing athletes to train harder. They also **increase aggression**.
- **Stimulants** — these drugs **speed up reactions**, **reduce fatigue** and **increase aggression**.
- **Narcotic analgesics** — these drugs **reduce pain**, so **injuries don't affect performance**.

Performance-enhancing drugs are **banned** in most sports. Athletes can be **tested** for drugs at any time and if they're **caught**, they can be **banned** from **competing** and stripped of any medals.

The *Use* of *Performance-Enhancing Drugs* is *Controversial*

There are many reasons why **performance-enhancing drugs** are **banned** but some people think they should be **allowed** in sport. You need to know **both sides** of the argument:

Arguments AGAINST using performance-enhancing drugs

- Some performance-enhancing drugs are **illegal**.
- Competitions become **unfair** if some people take drugs — people gain an advantage by taking drugs, not through training or hard work.
- There are some **serious health risks** associated with the drugs used, such as high blood pressure and heart problems.
- Athletes may **not** be **fully informed** of the health risks of the drugs they take.

Arguments FOR using performance-enhancing drugs

- It's up to each individual — athletes have the right to make their **own decision** about taking drugs and whether they're worth the risk or not.
- Drug-free sport **isn't** really **fair** anyway — different athletes have access to different training facilities, coaches, equipment, etc.
- Athletes that want to compete at a **higher level** may only be able to by using performance-enhancing drugs.

Practice Questions

Q1 Where are your cruciate ligaments?

Q2 What are performance-enhancing drugs?

Exam Questions

Q1 A cricketer suffers a knee injury during a game. When examined by doctors, he's told he may need surgery.
   a) Give three benefits of keyhole surgery compared to open surgery. [3 marks]
   b) The damage to his knee may be so bad that it can't be repaired. How could a prosthesis help? [2 marks]

Q2 Many sports organisations have banned the use of performance-enhancing drugs.
   Give two arguments in favour of such a ban. [2 marks]

Through the keyhole — who operates on a knee like this...

Wow, I didn't know sport could be so technical. Eeeee, if you hurt yourself when I were a lad, you'd be lucky to get yourself a sticky plaster... Anyway, it seems the world of sport has come on a bit since then. And you know what that means — yep, it's crucial you learn about cruciates and paramount you learn about prosthetics for your exam.

Nervous and Hormonal Communication

Right, it's time to get your brain cells fired up and get your teeth stuck into a mammoth — a mammoth section, that is...

Responding to their Environment Helps Organisms Survive

1) **Animals increase** their **chances** of **survival** by **responding** to **changes** in their **external environment**, e.g. by **avoiding harmful environments** such as places that are too hot or too cold.

2) They also **respond** to **changes** in their **internal environment** to make sure that the **conditions** are always **optimal** for their **metabolism** (all the chemical reactions that go on inside them).

3) **Plants** also **increase** their **chances** of **survival** by **responding** to **changes** in their **environment** (see p. 82).

4) Any **change** in the internal or external **environment** is called a **stimulus**.

Receptors Detect Stimuli and Effectors Produce a Response

1) **Receptors detect stimuli** — they can be **cells** or **proteins** on **cell surface membranes**. There are **loads** of **different types** of receptors that detect **different stimuli**.

2) **Effectors** are cells that bring about a **response** to a **stimulus**, to produce an **effect**. Effectors include **muscle cells** and cells found in **glands**, e.g. the **pancreas**.

3) Receptors **communicate** with effectors via the **nervous system** (see below) or the **hormonal system** (see the next page), or sometimes using **both**.

The Nervous System Sends Information as Electrical Impulses

1) The **nervous system** is made up of a **complex network** of cells called **neurones**. There are **three main types** of neurone:

   - **Sensory neurones** transmit electrical impulses from **receptors** to the **central nervous system (CNS)** — the **brain** and **spinal cord**.
   - **Motor neurones** transmit electrical impulses from the **CNS** to **effectors**.
   - **Relay neurones** transmit electrical impulses **between** sensory neurones and motor neurones.

   There's more about the different types of neurone on page 76.

2) A stimulus is detected by **receptor cells** and an **electrical impulse** is sent along a **sensory neurone**.

3) When an **electrical impulse** reaches the end of a neurone chemicals called **neurotransmitters** take the information across to the **next neurone**, which then sends an **electrical impulse** (see p. 80).

   Electrical impulses are also called nerve impulses.

4) The **CNS processes** the information and sends impulses along **motor neurones** to an **effector**.

5) You need to know how your **eyes respond** to **bright light** (to **protect** them) or **dim light** (to **help you see better**):

Stimulus	→	Receptors	—sensory neurone→	CNS	—motor neurone→	Effectors	→	Response
Dim light.		**Light receptors (photoreceptors)** in your eyes **detect** the lack of light.		CNS **processes** the **information**.		**Radial muscles** in the **iris** are stimulated by the motor neurones.		**Radial muscles contract** to **dilate** your **pupils** (make them bigger).

The brain unconsciously processes the information, so these responses are reflexes.

Stimulus	→	Receptors	—sensory neurone→	CNS	—motor neurone→	Effectors	→	Response
Bright light.		**Light receptors (photoreceptors)** in your eyes **detect** the bright light.		CNS **processes** the **information**.		**Circular muscles** in the **iris** are stimulated by the motor neurones.		**Circular muscles contract** to **constrict** your **pupils**.

Nervous and Hormonal Communication

The **Hormonal System** Sends Information as **Chemical Signals**

1) The **hormonal system** is made up of **glands** and **hormones**:

- A **gland** is a group of cells that are specialised to **secrete** a useful substance, such as a **hormone**. E.g. the **pancreas** secretes **insulin**.
- **Hormones** are 'chemical messengers'. Many hormones are **proteins** or **peptides**, e.g. **insulin**. Some hormones are **steroids**, e.g. **progesterone**.

2) **Hormones** are **secreted** when a **gland** is **stimulated**:

- Glands can be **stimulated** by a **change** in **concentration** of a specific **substance** (sometimes **another hormone**).
- They can also be **stimulated** by **electrical impulses**.

3) Hormones **diffuse directly into** the **blood**, then they're **taken** around the body by the **circulatory system**.

4) They **diffuse out** of the blood **all over** the **body** but each hormone will only **bind** to **specific receptors** for that hormone, found on the membranes of some cells (called **target cells**).

5) The hormones trigger a **response** in the **target cells** (the **effectors**).

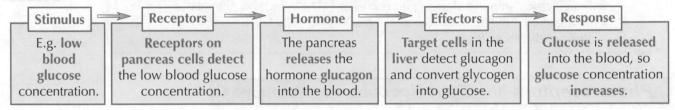

Stimulus		Receptors		Hormone		Effectors		Response
E.g. **low blood glucose** concentration.		**Receptors on pancreas cells detect** the low blood glucose concentration.		The pancreas **releases** the hormone **glucagon** into the blood.		**Target cells** in the liver detect glucagon and convert glycogen into glucose.		**Glucose** is **released** into the blood, so **glucose** concentration **increases**.

You Might Have to **Compare Nervous** and **Hormonal Communication**

Nervous Communication	Hormonal Communication
Uses electrical impulses.	Uses chemicals.
Faster response — electrical impulses are really fast.	Slower response — hormones travel at the 'speed of blood'.
Localised response — neurones carry electrical impulses to specific cells.	Widespread response — target cells can be all over the body.
Short-lived response — neurotransmitters are removed quickly.	Long-lived response — hormones aren't broken down very quickly.

Practice Questions

Q1 Why do organisms respond to changes in their environment?

Q2 Give two types of effector.

Q3 What is a hormone?

Exam Questions

Q1 Bright light causes circular iris muscles in an animal's eyes to contract, which constricts the pupils and protects the eyes. Describe and explain the roles of receptors and effectors in this response. [5 marks]

Q2 Describe three ways in which nervous communication is different from hormonal communication. [3 marks]

Vacancy — talented gag writer required for boring biology topics...

Actually, this stuff is really quite fascinating once you realise just how much your body can do without you even knowing. Just sit back and relax, let your nerves and hormones do the work... Ah, apart from the whole revision thing — your body can't do that without you knowing, unfortunately. Get your head around these pages before you tackle the rest of the section.

The Nervous System — Receptors

So now you know why organisms respond it's time for the (slightly less interesting but equally important) details... first up — receptors.

Receptors are Specific to One Kind of Stimulus

1) Receptors are **specific** — they only **detect one particular stimulus**, e.g. light, pressure or glucose concentration.

2) There are **many different types** of receptor that each detect a **different type of stimulus**.

3) Some receptors are **cells**, e.g. photoreceptors are receptor cells that connect to the nervous system. Some receptors are **proteins** on **cell surface membranes**, e.g. glucose receptors are proteins found in the cell membranes of some pancreatic cells.

4) When a nervous system receptor is in its **resting state** (not being stimulated), there's a **difference in charge** between the **inside** and the **outside** of the cell. This means there's a **voltage** across the membrane. The membrane is said to be **polarised**.

5) The **voltage** across the membrane is called the **potential difference**.

6) It is **generated** by **ion pumps** and **ion channels** (see p. 76).

7) When a **stimulus** is detected, the **permeability** of the **cell membrane** to **ions changes** (ions are stopped from moving, or more move in and out of the cell). This **changes the potential difference**.

8) If the **change** in potential difference is **big enough** it'll trigger an **action potential** — an electrical impulse along a neurone (see p. 77). An action potential is only triggered if the potential difference reaches a certain level called the **threshold** level.

Forget polarised membranes, it's all about polarised hair darling...

Photoreceptors are Light Receptors in Your Eye

1) **Light** enters the eye through the **pupil**. The **amount** of light that enters is **controlled** by the muscles of the **iris**.

2) Light rays are **focused** by the **lens** onto the **retina**, which lines the inside of the eye. The retina contains **photoreceptor cells** — these **detect light**.

3) The **fovea** is an area of the retina where there are **lots of photoreceptors**.

4) **Nerve impulses** from the photoreceptor cells are carried from the **retina** to the **brain** by the **optic nerve**, which is a bundle of **neurones**. Where the optic nerve leaves the eye is called the **blind spot** — there **aren't** any **photoreceptor cells**, so it's **not sensitive** to **light**.

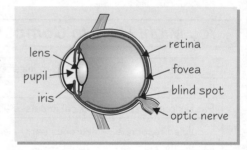

Photoreceptors Convert Light into an Electrical Impulse

1) **Light** enters the eye, hits the **photoreceptors** and is **absorbed** by **light-sensitive pigments**.

2) Light bleaches the pigments, causing a **chemical change**.

3) This triggers a **nerve impulse** along a **bipolar neurone**.

4) Bipolar neurones connect **photoreceptors** to the **optic nerve**, which takes impulses to the **brain**.

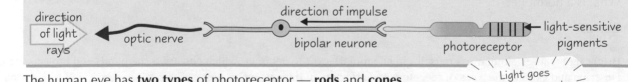

Light goes straight through the neurones to the photoreceptors.

5) The human eye has **two types** of photoreceptor — **rods** and **cones**.

6) Rods are mainly found in the **peripheral** parts of the **retina**, and cones are found **packed together** in the **fovea**.

7) Rods only give information in **black and white** (monochromatic vision), but cones give information in **colour** (trichromatic vision). There are three types of cones — **red-sensitive**, **green-sensitive** and **blue-sensitive**. They're stimulated in **different proportions** so you see different colours.

The Nervous System — Receptors

Rod Cells Hyperpolarise when Stimulated by Light

Rods contain a light-sensitive pigment called **rhodopsin**.
Rhodopsin is made of **two chemicals** joined together — **retinal** and **opsin**.
When it's **dark**, your rods **aren't stimulated** — here's what happens:

1) **Sodium ions** (Na^+) are **pumped out** of the cell using **active transport**.

2) But sodium ions **diffuse back in** to the cell through **open sodium channels**.

3) This makes the **inside** of the cell **only slightly negative** compared to the outside — the cell membrane is said to be **depolarised**.

4) This triggers the **release** of **neurotransmitters**.

5) But the neurotransmitters **inhibit** the **bipolar neurone** — the bipolar neurone **can't fire** an **action potential** so **no information** goes to the brain.

'Depolarised' means there's not much difference in charge across the membrane.

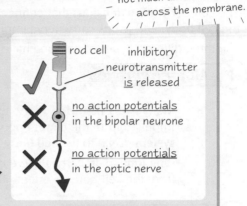

rod cell — inhibitory neurotransmitter <u>is</u> released

<u>no action potentials</u> in the bipolar neurone

<u>no action potentials</u> in the optic nerve

When it's **light**, your rod cells **are stimulated** — here's what happens:

1) **Light energy** causes **rhodopsin** to **break apart** into **retinal** and **opsin** — this process is called **bleaching**.

2) The bleaching of rhodopsin causes the **sodium ion channels** to **close**.

3) So **sodium ions** are actively transported **out** of the cell, but they **can't diffuse back in**.

4) This means sodium ions build up on the **outside** of the cell, making the **inside** of the membrane **much more negative** than the outside — the cell membrane is **hyperpolarised**.

5) When the rod cell is hyperpolarised it **stops releasing neurotransmitters**. This means there's **no inhibition** of the **bipolar neurone**.

6) Because the bipolar neurone is no longer inhibited, it **depolarises**. If the **change** in **potential difference** reaches the **threshold**, an **action potential** is transmitted to the **brain** via the **optic nerve**.

Sodium channels are cation channels because they only let positively charged ions (cations) through.

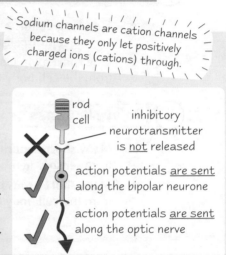

rod cell — inhibitory neurotransmitter is <u>not</u> released

action potentials <u>are sent</u> along the bipolar neurone

action potentials <u>are sent</u> along the optic nerve

Practice Questions

Q1 How many types of stimuli does one type of receptor detect?

Q2 Name the bundle of neurones that links the eye to the brain.

Q3 What do photoreceptors convert light into?

Q4 What do bipolar neurones connect?

Q5 When it's dark, is a bipolar neurone inhibited or uninhibited?

Exam Question

Q1 Explain how light falling on a rod cell triggers an action potential to be transmitted to the brain. [7 marks]

Rods — useful for both seeing and catching fish...

Wow, loads of stuff here, so cone-gratulations if you manage to remember it. In fact, get someone to test you, just to make sure it's well and truly fixed in that big grey blob you call your brain. Remember, receptors are really important because without them you wouldn't be able to see this book, and without this book revision would be way trickier.

The Nervous System — Neurones

Ah, on to the good stuff — how neurones carry info (in the form of action potentials) to other parts of the body...

You Need to **Learn** the **Structure** and **Function** of **Neurones**

1) All neurones have a **cell body** with a **nucleus** (plus **cytoplasm** and all the other **organelles** you usually get in a cell).

2) The cell body has **extensions** that **connect** to **other neurones** — dendrites carry nerve impulses **towards** the **cell body**, and **axons** carry nerve impulses **away** from the **cell body**.

3) The three different types of neurone have slightly different structures and different functions:

1 **Motor Neurones**

- **Many short dendrites** carry nerve impulses from the **central nervous system** (CNS) to the **cell body**.
- **One long axon** carries nerve impulses from the **cell body** to **effector cells**.

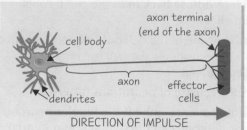

cell body
axon terminal (end of the axon)
axon
dendrites
effector cells

DIRECTION OF IMPULSE

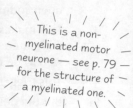
This is a non-myelinated motor neurone — see p. 79 for the structure of a myelinated one.

2 **Sensory Neurones**

- **One long dendrite** carries nerve impulses from **receptor cells** to the **cell body**.
- **One short axon** carries nerve impulses from the **cell body** to the **CNS**.

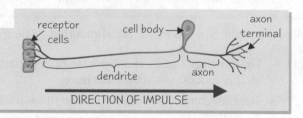

receptor cells
cell body
axon terminal
dendrite
axon

DIRECTION OF IMPULSE

3 **Relay Neurones**

- **Many short dendrites** carry nerve impulses from **sensory neurones** to the **cell body**.
- **Many short axons** carry nerve impulses from the **cell body** to **motor neurones**.

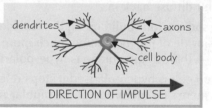

dendrites
axons
cell body

DIRECTION OF IMPULSE

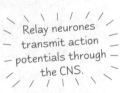
Relay neurones transmit action potentials through the CNS.

Neurone **Cell Membranes** are **Polarised** at **Rest**

1) In a neurone's **resting state** (when it's not being stimulated), the **outside** of the membrane is **positively charged** compared to the **inside**. This is because there are **more positive ions outside** the cell than inside.

2) So the membrane is **polarised** — there's a **difference in charge**.

3) The voltage across the membrane when it's at rest is called the **resting potential** — it's about **−70 mV**.

4) The resting potential is created and maintained by **sodium-potassium pumps** and **potassium ion channels** in a neurone's membrane:

Sodium-potassium pump

These pumps use **active transport** to move **three sodium ions** (Na^+) **out** of the neurone for every **two potassium ions** (K^+) moved **in**. ATP is needed to do this.

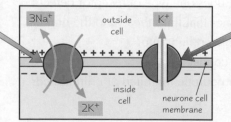

$3Na^+$
outside cell
K^+
inside cell
neurone cell membrane
$2K^+$

Potassium ion channel

These channels allow **facilitated diffusion** of **potassium ions** (K^+) **out** of the neurone, down their **concentration gradient**.

- The sodium-potassium pumps move **sodium ions out** of the neurone, but the membrane **isn't permeable** to **sodium ions**, so they **can't diffuse back in**. This creates a **sodium ion electrochemical gradient** (a **concentration gradient** of **ions**) because there are **more** positive sodium ions **outside** the cell than inside.

- The sodium-potassium pumps also move **potassium ions in** to the neurone, but the membrane **is permeable** to **potassium ions** so they **diffuse back out** through potassium ion channels.

- This makes the **outside** of the cell **positively charged** compared to the inside.

The Nervous System — Neurones

Neurone *Cell Membranes* Become *Depolarised* when they're *Stimulated*

A **stimulus** triggers other ion channels, called **sodium ion channels**, to **open**. If the stimulus is big enough, it'll trigger a **rapid change** in **potential difference**. The sequence of events that happen are known as an **action potential**:

① **Stimulus** — this **excites** the neurone cell membrane, causing **sodium ion channels** to **open**. The membrane becomes **more permeable** to sodium, so **sodium ions diffuse into** the neurone down the sodium ion electrochemical gradient. This makes the **inside** of the neurone **less negative**.

② **Depolarisation** — if the potential difference reaches the **threshold** (around **–55 mV**), **more sodium ion channels open**. **More sodium ions diffuse into** the neurone.

③ **Repolarisation** — at a potential difference of around **+30 mV** the **sodium ion channels close** and **potassium ion channels open**. The membrane is **more permeable** to potassium so **potassium ions diffuse out** of the neurone down the potassium ion concentration gradient. This starts to get the membrane **back** to its **resting potential**.

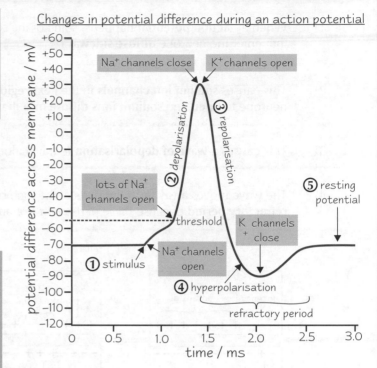

Changes in potential difference during an action potential

④ **Hyperpolarisation** — **potassium ion channels** are **slow to close** so there's a slight 'overshoot' where too many potassium ions diffuse out of the neurone. The potential difference becomes **more negative** than the **resting potential** (i.e. less than –70 mV).

The sodium and potassium channels are voltage-gated — they open at a certain voltage.

⑤ **Resting potential** — the ion channels are **reset**. The **sodium-potassium pump** returns the membrane to its **resting potential** and maintains it until the membrane's excited by another stimulus.

After an **action potential**, the neurone cell membrane **can't** be **excited** again straight away. This is because the ion channels are **recovering** and they **can't** be made to **open** — sodium ion channels are **closed** during repolarisation and **potassium ion channels** are **closed** during hyperpolarisation. This period of recovery is called the **refractory period**.

Practice Questions

Q1 Draw and label a motor neurone.
Q2 Name the pumps and channels that maintain a neurone's resting potential.

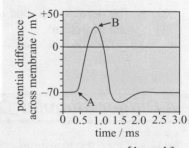

Exam Question

Q1 The graph shows an action potential across an axon membrane following the application of a stimulus.
  a) What label should be added at point A? [1 mark]
  b) Explain what causes the change in potential difference between point A and point B. [3 marks]
  c) A stimulus was applied at 1.5 ms, but failed to produce an action potential. Suggest why. [2 marks]

I'm feeling a bit depolarised after all that...

All this stuff about neurones can be a bit tricky to get your head around at first. Take your time and try scribbling it all down a few times till it starts to make some kind of sense. Neurones work because there's an electrical charge across their membrane, which is set up by ion pumps and ion channels. It's a change in this charge that transmits an action potential.

The Nervous System — Neurones

Action potentials don't just sit there once they've been generated — they have to hotfoot it all the way down the neurone so the information can be passed on to the next cell...

The **Action Potential** Moves **Along** the **Neurone** as a **Wave** of **Depolarisation**

1) When an **action potential** happens, some of the **sodium ions** that enter the neurone **diffuse sideways**.

2) This causes **sodium ion channels** in the **next region** of the neurone to **open** and **sodium ions diffuse into** that part.

3) This causes a **wave of depolarisation** to travel along the neurone.

4) The **wave** moves **away** from the parts of the membrane in the **refractory period** because these parts **can't fire** an action potential.

Cindy's wave activity was looking good...

It's like a Mexican wave travelling through a crowd — sodium ions rushing inwards causes a wave of activity along the membrane.

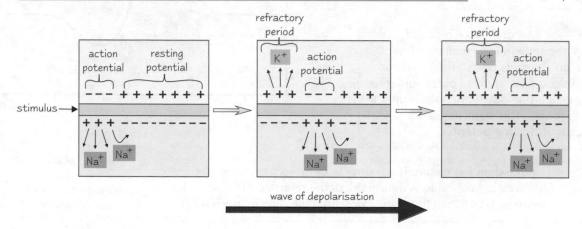

wave of depolarisation

The **Refractory Period** Produces **Discrete Impulses**

1) During the **refractory period**, **ion channels** are **recovering** and **can't** be **opened**.

2) So the refractory period acts as a **time delay** between one action potential and the next. This makes sure that **action potentials don't overlap** but pass along as **discrete** (separate) **impulses**.

3) The refractory period also makes sure **action potentials** are **unidirectional** (they only travel in **one direction**).

A **Bigger Stimulus** Causes **More Frequent Impulses**

1) Once the threshold is reached, an action potential will **always fire** with the **same change in voltage**, no matter how big the stimulus is.

2) If **threshold isn't reached**, an action potential **won't fire**.

3) A **bigger stimulus** won't cause a bigger action potential, but it will cause them to fire **more frequently**.

small stimulus

big stimulus

The Nervous System — Neurones

Action Potentials Go Faster in Myelinated Neurones

1) Some neurones are **myelinated** — they have a **myelin sheath**.
2) The myelin sheath is an **electrical insulator**.
3) It's made of a type of cell called a **Schwann cell**.
4) Between the Schwann cells are tiny patches of **bare membrane** called the **nodes of Ranvier**. **Sodium ion channels** are **concentrated** at the nodes.

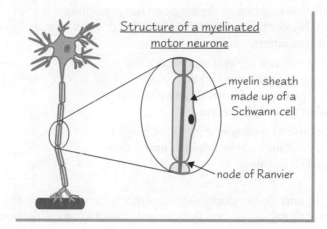

Structure of a myelinated motor neurone

myelin sheath made up of a Schwann cell

node of Ranvier

5) In a **myelinated** neurone, **depolarisation** only happens at the **nodes of Ranvier** (where sodium ions can get through the membrane).

6) The neurone's **cytoplasm conducts** enough electrical charge to **depolarise** the **next node**, so the impulse **'jumps'** from node to node.

7) This is called **saltatory conduction** and it's **really fast**.

8) In a **non-myelinated** neurone, the impulse travels as a **wave** along the **whole length** of the **axon membrane**.

9) This is **slower** than saltatory conduction (although it's still pretty quick).

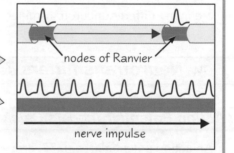

nodes of Ranvier

nerve impulse

Practice Questions

Q1 Briefly describe how an action potential moves along a neurone.
Q2 Give one function of the refractory period in nervous transmission.
Q3 How does a bigger stimulus affect the size of an action potential?
Q4 What is the function of Schwann cells on a neurone?
Q5 What are nodes of Ranvier?
Q6 Does a nerve impulse travel faster or slower along a non-myelinated neurone than a myelinated neurone?

Exam Question

Q1 Multiple sclerosis is a disease of the nervous system characterised by damage to the myelin sheaths of neurones. Explain how this will affect the transmission of action potentials. [5 marks]

Never mind the ion channels, I need to recover after this lot...

I'd expect animals like cheetahs, or even humans, to have super myelinated neurones so that nerve impulses are conducted really fast. But no, apparently we've been outdone by shrimps. That's right, top of the leader board so far are shrimps with their myelinated giant nerve fibre conducting impulses faster than 200 m/s. Blimey, they must get a lot done in a day.

The Nervous System — Synapses

When an action potential arrives at the end of a neurone the information has to be passed on to the next cell — this could be another neurone, a muscle cell or a gland cell.

A **Synapse** is a **Junction** Between a **Neurone** and the **Next Cell**

1) A **synapse** is the junction between a **neurone** and another **neurone**, or between a **neurone** and an **effector cell**, e.g. a muscle or gland cell.

2) The **tiny gap** between the cells at a synapse is called the **synaptic cleft**.

3) The **presynaptic neurone** (the one before the synapse) has a **swelling** called a **synaptic knob**. This contains **synaptic vesicles** filled with **chemicals** called **neurotransmitters**.

4) When an **action potential** reaches the end of a neurone it causes **neurotransmitters** to be **released** into the synaptic cleft. They **diffuse across** to the **postsynaptic membrane** (the one after the synapse) and **bind** to **specific receptors**.

5) When neurotransmitters bind to receptors they might **trigger** an **action potential** (in a neurone), cause **muscle contraction** (in a muscle cell), or cause a **hormone** to be **secreted** (from a gland cell).

6) Because the receptors are **only** on the postsynaptic membranes, synapses make sure **impulses** are **unidirectional** — the impulse can only travel in **one direction**.

7) Neurotransmitters are **removed** from the **cleft** so the **response** doesn't keep happening, e.g. they're taken back into the **presynaptic neurone** or they're **broken down** by **enzymes** (and the products are taken into the neurone).

8) There are many **different** neurotransmitters, e.g. **acetylcholine** and **dopamine**.

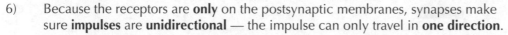

Typical structure of a synapse

presynaptic membrane

postsynaptic membrane

synaptic knob

vesicle filled with neurotransmitters

synaptic cleft

receptors

Here's How **Neurotransmitters Transmit Nerve Impulses Between Neurones**

1 An **Action Potential** Triggers **Calcium Influx**

1) An action potential (see p. 77) arrives at the **synaptic knob** of the **presynaptic neurone**.

2) The action potential stimulates **voltage-gated calcium ion channels** in the **presynaptic neurone** to **open**.

3) **Calcium ions diffuse into** the synaptic knob. (They're pumped out afterwards by active transport.)

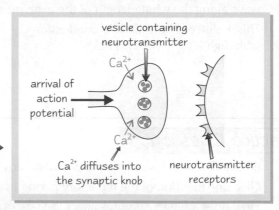

vesicle containing neurotransmitter

Ca^{2+}

arrival of action potential

Ca^{2+}

Ca^{2+} diffuses into the synaptic knob

neurotransmitter receptors

Synaptic knobs contain lots of mitochondria — they make ATP, which is needed for active transport and the movement of vesicles.

2 **Calcium Influx** Causes **Neurotransmitter Release**

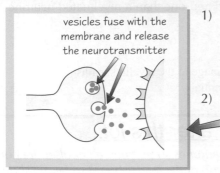

vesicles fuse with the membrane and release the neurotransmitter

1) The influx of **calcium ions** into the synaptic knob causes the **synaptic vesicles** to **move** to the **presynaptic membrane**. They then **fuse** with the presynaptic membrane.

2) The **vesicles release** the neurotransmitter into the **synaptic cleft** — this is called **exocytosis**.

The Nervous System — Synapses

(3) *The **Neurotransmitter Triggers** an **Action Potential** in the **Postsynaptic Neurone***

1) The neurotransmitter **diffuses** across the **synaptic cleft** and **binds** to specific **receptors** on the **postsynaptic membrane**.

2) This causes **sodium ion channels** in the **postsynaptic neurone** to **open**.

3) The **influx** of **sodium ions** into the postsynaptic membrane causes **depolarisation**. An **action potential** on the **postsynaptic membrane** is generated if the **threshold** is reached.

4) The **neurotransmitter** is **removed** from the **synaptic cleft** so the **response** doesn't keep happening.

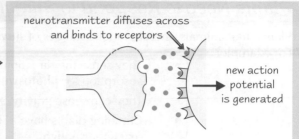

neurotransmitter diffuses across and binds to receptors

new action potential is generated

Synapses Play Vital Roles in the Nervous System

(1) **Synapses** allow **Information** to be **Dispersed** or **Amplified**

1) When **one** neurone **connects** to **many** neurones information can be **dispersed** to **different parts** of the body. This is called **synaptic divergence**.

2) When **many** neurones **connect** to **one** neurone information can be **amplified** (made stronger). This is called **synaptic convergence**.

Impulses diverge

Impulses converge

(2) **Summation** at **Synapses Finely Tunes** the **Nervous Response**

If a stimulus is **weak**, only a **small amount** of **neurotransmitter** will be released from a neurone into the synaptic cleft. This might not be enough to **excite** the postsynaptic membrane to the **threshold** level and stimulate an action potential. **Summation** is where the effect of neurotransmitter released from **many neurones** (or **one** neurone that's stimulated **a lot** in a short period of time) is **added together**.

Practice Questions

Q1 What is a synapse?

Q2 What name is given to the tiny gap between the cells at a synapse?

Q3 How do synapses ensure that nerve impulses are unidirectional?

Q4 Give one way that neurotransmitters are removed from the synaptic cleft.

Q5 Give one example of a neurotransmitter.

Q6 What is synaptic divergence?

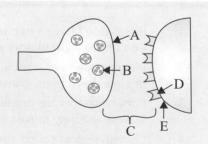

Exam Questions

Q1 The diagram on the right shows a synapse. Label parts A-E. [5 marks]

Q2 Describe the sequence of events from the arrival of an action potential at the presynaptic membrane of a synapse to the generation of a new action potential at the postsynaptic membrane. [6 marks]

<u>Synaptic knobs and clefts — will you stop giggling at the back...</u>

Some more pretty tough pages here, aren't I kind to you. And lots more diagrams to have a go at drawing and re-drawing. Don't worry if you're not the world's best artist, just make sure you add labels to your drawings to explain what's happening. Then there are only two more pages of this section before you can put your feet up and have a well-earned break.

Responses in Plants

Plants also have ways of responding to stimuli — OK so they're not as quick as animals, but they're important all the same.

Plants Need to Respond to Stimuli Too

Plants, like animals, **increase** their chances of **survival** by **responding** to changes in their **environment**. For example:

- They sense the direction of **light** and **grow** towards it to **maximise** light absorption for **photosynthesis**.
- They can sense **gravity**, so their roots and shoots **grow** in the **right direction**.
- **Climbing** plants have a sense of **touch**, so they can find things to climb and **reach** the **sunlight**.

A Tropism is a Plant's Growth Response to an External Stimulus

1) A **tropism** is the **response** of a plant to a **directional stimulus** (a stimulus coming from a particular direction).
2) Plants respond to directional stimuli by **regulating** their **growth**.
3) A **positive tropism** is growth **towards** the stimulus.
4) A **negative tropism** is growth **away** from the stimulus.

- **Phototropism** is the growth of a plant in response to **light**.
- **Shoots** are **positively phototropic** and grow **towards** light.
- **Roots** are **negatively phototropic** and grow **away** from light.

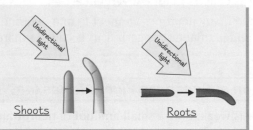

Shoots   Roots

- **Geotropism** is the growth of a plant in response to **gravity**.
- **Shoots** are **negatively geotropic** and grow **upwards**.
- **Roots** are **positively geotropic** and grow **downwards**.

The men's gymnastics team were negatively geotropic.

Responses are Brought About by Growth Factors

1) Plants **don't** have a **nervous system** so they can't respond using neurones, and they **don't** have a **circulatory system** so they can't respond using hormones either.
2) Plants **respond** to stimuli using **growth factors** — these are chemicals that **speed up** or **slow down** plant **growth**.
3) Growth factors are **produced** in the **growing regions** of the plant (e.g. shoot tips, leaves) and they **move** to where they're needed in the **other parts** of the plant.
4) Growth factors called **auxins** stimulate the **growth** of shoots by **cell elongation** — this is where **cell walls** become **loose** and **stretchy**, so the cells get **longer**.
5) **High** concentrations of auxins **inhibit growth** in **roots** though.
6) There are many **other** plant growth factors such as:
   - **Gibberellins** — stimulate **flowering** and **seed germination**.
   - **Cytokinins** — stimulate **cell division** and **cell differentiation**.
   - **Ethene** — stimulates **fruit ripening** and **flowering**.
   - **Abscisic acid (ABA)** — involved in **leaf fall**.

Responses in Plants

Indoleacetic Acid (IAA) is an Important Auxin

1) **Indoleacetic acid (IAA)** is an important **auxin** that's produced in the **tips** of **shoots** in flowering plants.

2) IAA is **moved** around the plant to **control tropisms** — it moves by **diffusion** and **active transport** over short distances, and via the **phloem** over longer distances.

3) This results in **different parts** of the plants having **different amounts** of IAA.
The **uneven distribution** of IAA means there's **uneven growth** of the plant, e.g:

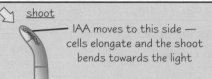

Phototropism — IAA moves to the more **shaded** parts of the **shoots** and **roots**, so there's uneven growth.

shoot — IAA moves to this side — cells elongate and the shoot bends towards the light

IAA moves to this side — growth is inhibited so the root bends away from the light — root

Geotropism — IAA moves to the **underside** of **shoots** and **roots**, so there's uneven growth.

shoot — IAA moves to this side — cells elongate so the shoot grows upwards

IAA moves to this side — growth is inhibited so the root grows downwards — root

Plants Detect Light Using Photoreceptors

1) Plants **detect light** using **photoreceptors** called **phytochromes**.

2) They're **found** in **many parts** of a plant including the **leaves**, **seeds**, **roots** and **stem**.

3) Phytochromes **control** a range of **responses**. For example, plants **flower** in **different seasons** depending on how much **daylight** there is at that time of year, e.g. some plants flower during **summer** when there are **long days**.

4) Phytochromes are **molecules** that **absorb light**. They exist in **two states** — the P_R state absorbs **red light** at a wavelength of **660 nm**, and the P_{FR} state absorbs **far-red light** at a wavelength of **730 nm**.

5) Phytochromes are **converted** from **one state** to **another** when **exposed** to **light**:
   - P_R is **quickly converted** into P_{FR} when it's exposed to **red light**.
   - P_{FR} is **quickly converted** into P_R when it's exposed to **far-red light**.
   - P_{FR} is **slowly converted** into P_R when it's in **darkness**.

P_R — red light (fast) → P_{FR}
← far-red light (fast)
darkness (slow)

6) **Daylight** contains **more red light** than **far-red light**, so **more** P_R is converted into P_{FR} than P_{FR} is converted to P_R.

7) So the **amount** of P_R and P_{FR} **changes** depending on the **amount of light**, e.g. whether it's **day** or **night**, or **summer** or **winter**.

8) It's the **differing amounts of** P_R **and** P_{FR} that **control** the **responses** to **light**. E.g. flowering — in some plants **high levels** of P_{FR} **stimulates flowering**. When **nights** are **short** in the summer, there's not much time for P_{FR} to be converted back into P_R, so P_{FR} **builds up**. This means the plants **flower** during the **summer**.

Practice Questions

Q1 Why do plants need to respond to stimuli?

Q2 What is a tropism?

Q3 What are plant growth factors?

Exam Questions

Q1 Explain how the movement of IAA in a growing shoot enables the plant to grow towards the light. [3 marks]

Q2 Iris plants are stimulated to flower by high levels of P_{FR}.
a) What is P_{FR}? [1 mark]
b) Suggest what time of year an iris would flower and explain your answer. [4 marks]

Auxin Productions — do you have the growth factor — with Simon Trowel...

I never knew plants were so complicated — you'd never guess, just by looking at one. I thought they just grew a bit, flowered if you're lucky and then died (mine mostly seem to die — I definitely don't have green fingers). Make sure you learn these pages about growth factors, get your head around a plant's bendy responses and then you've done the section.

Brain Structure and Function

So... the brain... a big old squelchy mass that controls all the goings on in your body, from a sniffle to your heart rate. Without your brain you wouldn't be able to see, hear, think or learn — in fact, you wouldn't be much use at all...

Different Areas of the Brain control Different Functions

Your brain **controls** the rest of your body, but **different parts** control **different functions**. You need to know the **location** and **function** of these **four** brain structures:

① The Cerebrum Allows You to See, Think, Learn and Feel Emotions

1) The **cerebrum** is the **largest** part of the brain.

2) It's divided into **two halves** called **cerebral hemispheres**.

3) The cerebrum has a thin **outer layer** called the **cerebral cortex**. The cortex has a **large surface area** so it's **highly folded** to fit into the skull.

4) The cerebrum is involved in **vision**, **learning**, **thinking** and **emotions**.

5) **Different parts** of the **cerebrum** are involved in **different functions**, e.g. the **back** of the cortex is involved in **vision** and the **front** is involved in **thinking**.

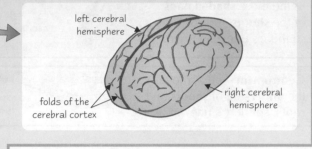

left cerebral hemisphere

folds of the cerebral cortex

right cerebral hemisphere

This diagram shows the brain cut in half from front to back — you might get a diagram of the brain from a different angle in your exam.

② The Hypothalamus Controls Body Temperature

1) The hypothalamus is found just **beneath** the **middle part** of the brain.

2) The hypothalamus automatically **maintains body temperature** at the normal level (**thermoregulation**) — see pages 66-67.

3) The hypothalamus produces **hormones** that **control** the **pituitary gland** — a gland just below the hypothalamus.

FRONT

BACK

pituitary gland

③ The Medulla Controls Breathing Rate and Heart Rate

1) The **medulla** is at the **base** of the **brain**, at the top of the spinal cord.

2) It automatically controls **breathing rate** and **heart rate**.

The medulla's proper name is the medulla oblongata.

④ The Cerebellum Coordinates Movement

1) The **cerebellum** is **underneath** the **cerebrum** and it also has a **folded cortex**.

2) It's important for **coordinating movement** and **balance**.

Come on cerebellum, don't fail me now, it's going to really hurt.

Brain Structure and Function

Scanners are used to Visualise the Brain

1) To investigate the **structure** and **function** of the brain, and to **diagnose medical conditions**, you need to **look inside** the brain.

2) This can be done with **surgery**, but it's **pretty risky**.

3) The brain can be visualised without surgery using **scanners**.

4) You need to **know** about the **three** different types of scanner:

1 Computed Tomography (CT) Scanners use Lots of X-rays

CT scanners use **radiation** (**X-rays**) to produce **cross-section images** of the brain. **Dense structures** in the brain **absorb more radiation** than less dense structures so show up as a **lighter colour** on the scan.

> Computers can build up many 2D images to produce a 3D image of the brain.

Uses of CT

Investigating brain structure — A CT scan shows the **major structures** in the brain.

Investigating brain function — A CT scan **doesn't** show brain **function** — it only shows brain structure. But if a CT scan shows a **diseased** or **damaged brain structure** and the patient has **lost some function**, the **function** of that part of the brain can be **worked out**. E.g. if an area of the brain is damaged and the patient can't see, then that area is involved in vision.

Medical diagnosis — CT scans can be used to **diagnose medical problems** because they show **damaged** or **diseased** areas of the brain, e.g. **bleeding** in the brain after a **stroke**:

- **Blood** has a **different density** from brain tissue so it shows up as a **lighter colour** on a CT scan.

- A scan will show the **extent** of the bleeding and its **location** in the brain.

- You can then work out **which blood vessels** have been damaged and what **brain functions** are likely to be **affected** by the bleeding.

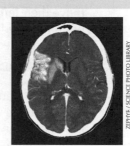

This CT scan looks up at the brain from below. The white area is heavy bleeding due to a blood vessel ruptured after a stroke.

ZEPHYR / SCIENCE PHOTO LIBRARY

2 Magnetic Resonance Imaging (MRI) Scanners use Magnetic Fields

MRI scanners use a **really strong magnetic field** and **radio waves** to produce **cross-section images** of the brain.

> An MRI scanner costs a <u>lot</u> more than a CT scanner.

Uses of MRI

Investigating brain structure — You can see the structure of the brain in a **lot more detail** using an MRI scanner than you can with a CT scanner, and you can **clearly see** the **difference** between **normal** and **abnormal** (diseased or damaged) brain tissue. For example, a scan can show diseased tissue caused by multiple sclerosis (a disease of the central nervous system).

Investigating brain function — This is done in the same way as with a CT scan.

Medical diagnosis — MRI scans can also be used to **diagnose medical problems** because they show **damaged** or **diseased** areas of the brain, e.g. a **brain tumour** (an abnormal mass of cells in the brain):

- **Tumour cells respond differently** to a **magnetic field** than healthy cells, so they show up as a **lighter colour** on an MRI scan.

- A scan will show the **exact size** of a tumour and its **location** in the brain. Doctors can then use this information to decide the most effective treatment.

- You can also work out what **brain functions** may be **affected** by the tumour.

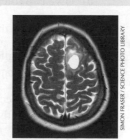

This MRI scan looks down on the brain from above. The white area is a tumour.

SIMON FRASER / SCIENCE PHOTO LIBRARY

Brain Structure and Function

3 Functional Magnetic Resonance Imaging (fMRI) Scanners show Brain Activity

fMRI scanners are **like MRI** scanners (see previous page), but they show **changes** in **brain activity** as they actually happen:

1) **More oxygenated blood** flows to **active areas** of the brain (to supply the neurones with oxygen and glucose).

2) Molecules in **oxygenated blood respond differently** to a **magnetic field** than those in deoxygenated blood.

3) So the **more active areas** of the brain can be **identified** on an fMRI scan.

Uses of fMRI

Investigating brain structure — An fMRI scan gives a **detailed** picture of the **brain's structure**, similar to an MRI scan.

Investigating brain function — fMRI scans are used to research the **function** of the brain as well as its structure. If a function is **carried out** whilst **in the scanner**, the **part** of the brain that's involved with that function will be **more active**. E.g. a patient might be asked to **move** their **left hand** when in the fMRI scanner. The areas of the brain involved in moving the hand will **show up** on the fMRI scan (these are often coloured by the computer so they show up more easily).

Medical diagnosis — fMRI scans are really useful to **diagnose medical problems** because they show **damaged** or **diseased areas** of the brain and they allow you to study conditions caused by **abnormal activity** in the brain (some conditions don't have an obvious structural cause). E.g. an fMRI scan can be taken of a patient's brain **before** and **during** a **seizure**. This can help to pinpoint which part of the brain's **not working properly** and find the **cause** of the seizure. Then the patient can receive the **most effective treatment** for the seizures.

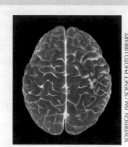

SOVEREIGN, ISM / SCIENCE PHOTO LIBRARY

This fMRI scan looks down on the brain from above. The red area is active when the person moves their left hand. The right side's active because it controls the left side of your body.

Practice Questions

Q1 Which part of the brain is involved in learning?

Q2 Which part of the brain controls body temperature?

Q3 What type of scanner produces an image of the brain using X-rays?

These questions cover pages 84-86.

Exam Questions

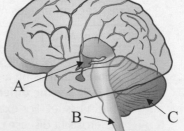

Q1 a) Name structure A on the diagram of the brain. [1 mark]

   b) Give two roles of structure B. [2 marks]

   c) What effect might damage to structure C have on the body? [1 mark]

Q2 A patient has fallen and hit his head. His doctor recommends an MRI (magnetic resonance imaging) scan to investigate suspected bleeding in his brain.

   a) Give two pieces of information about the bleeding that the doctor would be able to get from the MRI scan. [2 marks]

   b) Before operating on him, the surgeon wants to use a different scanner to assess his brain activity. Suggest what kind of scanner should be used. [1 mark]

The cere-mum part of the brain — coordinates dirty washing and clean clothes...

Ah, the good old brain. It's a mysterious creature alright, but the invention of scanners means scientists are starting to unlock its secrets. All things in the brain have hard-to-spell names unfortunately, like hypothalamus and medulla. Annoyingly cerebrum and cerebellum sound pretty similar — so make sure you learn that the <u>longer</u> name is at the <u>bottom</u> of the brain.

Brain Development and Habituation

I bet you've never sat down and had a good think about why your brain develops the way it does. Well now's the time...

You can Investigate the Role of **Nature and Nurture** in **Brain Development**

1) **Brain development** is how the brain **grows** and how **neurones connect together**. Measures of brain development include the **size** of the brain, the **number of neurones** it has and the **level of brain function** (e.g. speech, intelligence) a person has.

2) Your brain develops the way it does because of your **genes** (**nature**) and your **environment** (**nurture**) — your brain would **develop differently** if you had **different genes** or were brought up in a **different environment**.

3) Nature and nurture are **both involved** in controlling brain development, but **scientists disagree** about which one **influences** brain development the **most** — this **argument** is called the **nature-nurture debate**.

4) It's **really hard** to **investigate** the effects of **nature** and **nurture** because:

- Genetic and environmental factors interact, so it's difficult to know which one is the **main influence**.
- There are lots of different genes and lots of different environmental factors to investigate.
- To do an accurate experiment, you need to cancel out one factor to be able to investigate the other. This is really difficult to do — you'd need to cut out all environmental influences to investigate the role of a genetics, and vice versa.

5) You need to know the following **five methods** that are used to **investigate** the effects of **nature** and **nurture** on **brain development**:

1 Animal Experiments

1) Scientists study the effects of **different environments** (nurture) on the **brain development** of animals of the **same species**. Any differences in their brain development are **more likely** to be due to **nurture** than nature (if they're the same species they'll be genetically similar).

2) For example, animal experiments have shown that:

- Rats raised in a **stimulating environment** have **larger brains** and get **better scores** on **problem-solving tasks** than rats raised in boring environments (e.g. in a bare, dark cage). This suggests **nurture** plays a big role in **brain size** and the development of **problem-solving skills**.
- Rats reared in **isolation** (having no contact with other rats) have **similar brain abnormalities** to those found in **schizophrenic** patients. This suggests **nurture** plays a big role in **brain development**.

3) Scientists also study the effects of **different genes** (nature) on the **brain development** of animals raised in **similar environments**. They usually do this by **genetically engineering mice** to **lack** a particular **gene**. Any differences between the brain development of the genetically engineered mice and **normal mice** are **more likely** to be due to **nature** than nurture.

4) For example, animal experiments have shown that:

Mice engineered to **lack** the **Lgl1 gene** develop **enlarged brain regions** and **fluid builds up** in their brains. This suggests that **nature** plays a big role in **brain development**.

2 Newborn Studies

The brain of a newborn baby has been affected a bit by the environment in the womb.

1) The brain of a newborn baby **hasn't** really been **affected** by the **environment**.

2) Scientists study the brains of newborn babies to see what **functions** they're **born with** and **how developed different parts** of the brain are — what they're born with is **more likely** to be due to **nature** than nurture.

3) For example, newborn studies have shown that:

- Babies are **born** with a number of **abilities**, e.g. they can **cry**, **feed**, **recognise** a **human face**. This suggests that **nature** plays a big role in controlling these abilities.
- Newborn babies **don't** have the ability to **speak**, suggesting that **nurture** plays a big role in the ability to **speak**.

Brain Development and Habituation

3) Twin Studies

1) **Identical twins** are **genetically identical**.

2) If identical twins are **raised separately** then they'll have **identical genes** but **different environments**.

3) Scientists can compare the brain development of **separated identical twins** — any **differences** between them **are due to nurture** not nature, and any **similarities** between them are due to **nature**, for example:

   > Identical twins have **very similar IQ scores** — suggesting **nature** plays a big role in **intelligence**.

4) Scientists also **study** the brain development of **identical twins raised together**. These twins are **genetically identical** and have **similar environments**, so it's hard to tell if any differences between them are due to nature or nurture. So scientists compare them to **non-identical twins** (who are **genetically different** but have **similar environments**) — they act like a control to **cancel out** the **influence** of the **environment**. So any difference in brain development between identical and non-identical twins is **more likely** to be due to **nature** than nurture.

5) For example:
   - **Stuttering** of both twins is **more common** in **identical twins** than in non-identical twins. This suggests **nature** plays a big role in developing the **speech area** of the brain.
   - There's **no difference** in **reading ability** between pairs of identical and non-identical twins. This suggests **nurture** plays a big role in **reading ability**.

4) Brain Damage Studies

1) **Damage** to an adult's brain can lead to the **loss** of **brain function**, e.g. a stroke may cause loss of vision.

2) If an **adult's brain** is damaged, it can't repair itself so well because it's **already fully developed**. But a **child's brain** is **still developing** — so scientists can **study** the effects of **brain damage** on their development.

3) To do this, scientists **compare** the development of a chosen **function** (e.g. language) in children **with** brain damage to those **without** brain damage.

4) If the **characteristic still develops** in children who have brain damage, then brain development is **more likely** to be due to **nurture** than nature for that characteristic.

5) If the **characteristic doesn't develop** in children who have brain damage, then brain development is **more likely** to be due to **nature** than nurture for that characteristic (because nurture isn't having an effect).

6) For example:
   - Children aged 1-3 who were **born** with **damage** to the area of the brain associated with **language**, show a **delay** in the major language milestones (e.g. understanding words, producing sentences) when compared to children born without brain damage.
   - But by **age 5**, their **language skills** are the **same** as children with no brain damage — showing that if a young child's brain is damaged, they can **still develop language**.
   - This suggests that **nurture** plays a big role in **language development**.

5) Cross-Cultural Studies

1) Children brought up in **different cultures** have **different environmental influences**, e.g. social practices, beliefs, education, gender influences.

2) Scientists can study the **effects** of a different upbringing on **brain development** by comparing **large groups** of children who are the **same age** but from **different cultures**.

3) Scientists look for **major differences in characteristics** — any **differences** in brain development between different cultures are **more likely** to be due to **nurture** than nature. Any **similarities** in brain development between different cultures are **more likely** to be due to **nature** than nurture.

4) For example:
   - **Kenyan children** who eat **protein-rich food** (providing nutrients such as **zinc** and **iron**) have **higher IQs** than children who have a **poor diet** and **limited protein**. This suggests **nurture** plays a big role in **intelligence**.
   - The **mapping abilities** (e.g. perspective drawing) of young children are **well-developed across cultures**. This suggests that **nature** plays a big role in **mapping abilities**.

Brain Development and Habituation

Habituation is a type of Learned Behaviour

1) Animals (including humans) **increase** their chance of **survival** by **responding** to **stimuli** (see p. 72).

2) But if the stimulus is **unimportant** (if it's not threatening or rewarding), there's **no point** in **responding** to it.

3) If an unimportant stimulus is **repeated** over a period of **time**, an animal **learns** to **ignore it**.

   E.g. you **learn** to **sleep through traffic noise** at night.

4) This **reduced response** to an **unimportant stimulus** after **repeated** exposure **over time** is called **habituation**.

5) Habituation means animals **don't waste time** and **energy** responding to unimportant stimuli.

   E.g. **prairie dogs** use **alarm calls** to warn others of a predator — but they've **habituated** to **humans** because we're **not a threat**. They **no longer** send out alarm calls when they see humans, so they **don't waste time** or **energy**.

6) Animals still remain **alert** to **important stimuli** though (stimuli which might **threaten** their **survival**).

   E.g. you can sleep through traffic noise, but you'll **instantly wake up** if you hear an **unfamiliar noise**.

7) You need to know how to **investigate habituation** to a stimulus:

> To **investigate habituation**, you need to be able to **measure** an animal's **response** to an **unimportant stimulus**. Here's how you could do it with **tortoises**, but the principles are the **same** for any organism:
>
> 1) **Gently tap a tortoise** on its shell — it should cause the tortoise to **withdraw** its head, feet and tail **into its shell**.
>
> 2) **Time** how **long** it takes for the tortoise to **reappear** out of its shell after you've tapped it.
>
> 3) Tap the tortoise's shell at **regular intervals** (e.g. every minute) and **record** the **time** it takes for it to reappear.
>
>
> *Fearless Mr Tappy made the experiment very boring for Bob.*
>
> If **habituation has taken place** the tortoise should **reappear** out of its shell **quicker** the **more** you **repeat** the **stimulus** (or it **might not withdraw** at all eventually). If habituation **hasn't occurred** the tortoise will take the **same length of time** to **reappear** each time.
>
> The tortoise should still **remain alert** to an **unfamiliar stimulus**, e.g. if you **clap** your hands near the tortoise it should **cause it to withdraw** into its shell.

Practice Questions

Q1　What is meant by the term 'brain development'?

Q2　What is meant by the term 'nurture'?

These questions cover pages 87-89.

Exam Questions

Q1　A scientist carries out a series of studies on newborn babies to find out how brain development is affected by nature.

　　a) Explain what is meant by 'nature'. [1 mark]

　　b) What is the advantage of using newborn babies to study the effects of nature on brain development? [2 marks]

　　c) Suggest two other types of study that a scientist could carry out to directly research the effect of nature and nurture on brain development in humans. [2 marks]

Q2　A birdwatcher sat in his garden quietly for an hour each day. At first this scared the birds away, but over the next few weeks he saw more and more birds. The birdwatcher predicted that this was due to habituation.

　　a) Explain why the birds' behaviour could be described as habituation. [3 marks]

　　b) What is the advantage of habituation to birds in the wild? [1 mark]

You say nature and I say nurture — let's call the whole thing off...

Whether a brain function is influenced by nature or nurture is pretty important — if it's heavily influenced by nurture then you can figure out how to improve that brain function by changing an organism's environment. Clever stuff.

Development of the Visual Cortex

Your brain continues to develop even after you're born — good job too, you'd have trouble learning this stuff if it didn't.

Animal Models are used to Study Brain Development

Some animals have fairly **similar brains** to **humans**. This means scientists can do **experiments** on these animals (that would be **unethical** to do **in humans**) to **investigate brain development**, e.g. experiments have been done on cats to investigate the development of the visual cortex (see below).

The Visual Cortex is Made Up of Ocular Dominance Columns

1) The **visual cortex** is an area of the **cerebral cortex** (see p. 84) at the **back** of your brain.

2) The role of the visual cortex is to **receive** and **process visual information**.

3) **Neurones** in the visual cortex **receive information** from **either** your **left** or **right eye**.

4) Neurones are **grouped together** in columns called **ocular dominance columns**. If they receive information from the **right eye** they're called **right** ocular dominance columns, and if they receive information from the **left eye** they're called **left** ocular dominance columns.

5) The columns are the **same size** and they're arranged in an **alternating pattern** (left, right, left, right) across the visual cortex.

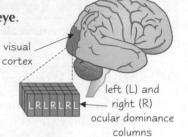

BACK   FRONT

visual cortex

left (L) and right (R) ocular dominance columns

Hubel and Wiesel Investigated How the Visual Cortex Develops

1) The **structure** of the **visual cortex** was **discovered** by two scientists called **Hubel** and **Wiesel**.

2) They used **cats** and **monkeys** to study the **electrical activity** of **neurones** in the visual cortex.

3) They found that the **left ocular dominance columns** were **stimulated** when an animal used its **left eye**, and the **right ocular dominance columns** were **stimulated** when it used its **right eye**.

4) Hubel and Wiesel investigated **how** the **visual cortex develops** by experimenting on very **young kittens**:

- They **stitched shut one eye** of each kitten so they could only see out of their other eye.
- The kittens were kept like this for **several months** before their eyes were **unstitched**.
- Hubel and Wiesel found that the kitten's **eye** that had been **stitched up** was **blind**.
- They also found that **ocular dominance columns** for the **stitched up eye** were a **lot smaller** than normal, and the ocular dominance columns for the **open eye** were a **lot bigger** than normal.
- The ocular dominance columns for the **open eye** had **expanded** to **take over** the other columns that **weren't** being stimulated — when this happens, the **neurones** in the visual cortex are said to have **switched dominance**.

5) Hubel and Wiesel then investigated if the **same thing** happened in an **adult's brain**, e.g. they experimented on **cats**:

- They **stitched shut one eye** of each cat, who were kept like this for **several months**.
- When their eyes were **unstitched**, Hubel and Wiesel found that these eyes **hadn't gone blind**.
- The cats **fully recovered** their **vision** and their **ocular dominance columns** remained the **same**.

6) Hubel and Wiesel **repeated** the experiments on **young monkeys** and **adult monkeys** — they saw the **same results**.

7) Hubel and Wiesel's experiments showed that the **visual cortex only develops** into normal **left** and **right** ocular dominance columns if **both eyes** are **visually stimulated** in the **very early stages** of **life**.

Their Experiments Provide Evidence for a Critical 'Window' in Humans

1) Hubel and Wiesel's experiments on cats show there's a **period** in **early life** when it's **critical** that a kitten is **exposed** to **visual stimuli** for its visual cortex to **develop properly**. This period of time is called the **critical 'window'**.

2) The **human visual cortex** is **similar** to a **cat's visual cortex** (the human visual cortex has **ocular dominance columns too**) so Hubel and Wiesel's experiments provide **evidence** for a **critical 'window'** in **humans**.

3) There's **other evidence** for a visual cortex critical 'window' in humans as well (see next page).

Development of the Visual Cortex

Evidence from Human Studies Suggests a Critical 'Window' Does Exist

Scientists have **investigated** how the **visual system develops** in **humans**, e.g. by looking at **cataracts** in the **eye**:

- A **cataract** makes the **lens** in the eye go **cloudy**, causing **blurry vision**.
- If a **baby** has a **cataract**, it's **important** to **remove** the cataract within the **first few months** of the baby's life — otherwise their visual system **won't develop properly** and their vision will be **damaged for life**.
- If an **adult** has a **cataract** then it's not so serious — when the cataract is **removed**, **normal vision** comes back **straight away**. This is because the visual system is **already developed** in an adult.

Using Animals in Medical Research Raises Ethical Issues

Hubel and Wiesel used **animals** in their experiments, which is **common** in **medical research**. This raises some **ethical issues** — you need to **know** a range of **arguments** from **both sides**:

Arguments AGAINST using animals in medical research	Arguments FOR using animals in medical research
Animals are different from humans, so drugs tested on animals may have different effects in humans.	Animals are similar to humans, so research has led to loads of medical breakthroughs, e.g. antibiotics, insulin for diabetics, organ transplants.
Experiments can cause pain and distress to animals.	Animal experiments are only done when it's absolutely necessary and scientists follow strict rules, e.g. animals must be properly looked after, painkillers and anaesthetics must be used to minimise pain.
There are alternatives to using animals in research, e.g. using cultures of human cells or using computer models to predict the effects of experiments.	Using animals is currently the only way to study how a drug affects the whole body — cell cultures and computers aren't a true representation of how cells may react when surrounded by other body tissues. It's also the only way to study behaviour.
Some people think that animals have the right to not be experimented on, e.g. animal rights activists.	Other people think that humans have a greater right to life than animals because we have more complex brains, e.g. compared to rats, fish, fruit flies (which are commonly used in experiments).

Practice Questions

Q1 Where in the brain are ocular dominance columns found?

Q2 What kind of pattern are ocular dominance columns arranged in?

Q3 Describe one piece of evidence that suggests a critical window exists for human visual system development.

Q4 Give two arguments against the use of animals in medical research.

Exam Question

Q1 Hubel and Wiesel conducted experiments on animals to investigate the structure and development of the visual cortex.

a) Describe their experiment on kittens and explain what this showed about how the visual cortex develops. [4 marks]

b) Do their experiments give evidence for a critical 'window' in the development of the human visual system? Explain your answer, with reference to what is meant by a critical 'window'. [2 marks]

c) Give two arguments for using animals in medical research. [2 marks]

Hubel and Wiesel — weren't they a '60s pop duo...

The experiments that Hubel and Wiesel did on cats and monkeys may have been a bit gross, but they did provide us with good knowledge of how the visual cortex develops. There's evidence to suggest a critical 'window' for other things too, like language development. Interesting stuff, so make sure you learn it and then you can impress others (like the examiners).

Drugs and Disease

Brace yourself — this section isn't as exciting as the name suggests, but it is the last. Let revision commence...

Imbalances in Some Neurotransmitters can Contribute to Disorders

Neurotransmitters are **chemicals** that **transmit** nerve impulses across **synapses** (see page 80). Some **disorders** are **linked** to an **imbalance** of specific neurotransmitters in the brain. Here are two examples you need to know:

Parkinson's Disease

1) Parkinson's disease is a **brain disorder** that affects the **motor skills** (the movement) of people.

2) In Parkinson's disease the **neurones** in the **parts** of the **brain** that **control movement** are **destroyed**.

3) These neurones **normally produce** the neurotransmitter **dopamine**, so **losing them** causes a **lack** of **dopamine**.

4) A lack of dopamine causes a **decrease** in the **transmission** of the **nerve impulses** involved in **movement**.

5) This leads to **symptoms** like **tremors** (shaking) and **slow movement**.

6) Scientists know that the **symptoms** are **caused by** a **lack** of **dopamine** so they've **developed drugs** (e.g. **L-dopa**, see below) to **increase** the level of **dopamine** in the brain.

Depression

1) Scientists think there's a **link** between a **low level** of the neurotransmitter **serotonin** and **depression**.

2) Serotonin transmits **nerve impulses** across synapses in the **parts** of the **brain** that **control mood**.

3) Scientists know that **depression** is **linked** to a **low level** of serotonin so they've **developed drugs** (antidepressants) to **increase** the level of **serotonin** in the brain.

Some Drugs Work by Affecting Synaptic Transmission

See p. 80 for more on synaptic transmission.

You need to know these two examples:

L-dopa

1) L-dopa is a drug that's used to **treat** the **symptoms** of **Parkinson's disease**.

2) Its **structure** is very **similar** to **dopamine**.

3) When L-dopa is given, it's **absorbed** into the **brain** and **converted** into **dopamine** by the enzyme **dopa-decarboxylase** (dopamine can't be given to treat Parkinson's disease because it **can't enter** the **brain**).

4) This **increases** the level of **dopamine** in the brain.

5) A higher level of dopamine means that **more nerve impulses** are **transmitted** across synapses in the **parts** of the **brain** that **control movement**.

6) This gives sufferers of Parkinson's disease **more control** over their **movement**.

MDMA (ecstasy)

1) MDMA **increases** the level of **serotonin** in the brain.

2) Usually, serotonin is **taken back** into a **presynaptic neurone** after triggering an action potential, to be **used again**.

3) MDMA **increases** the level of **serotonin** by **inhibiting** the **reuptake** of serotonin **into presynaptic neurones**, and by **triggering** the **release** of serotonin **from presynaptic neurones**.

4) This means that **nerve impulses** are **constantly triggered** in postsynaptic neurones in **parts** of the **brain** that **control mood**.

5) So the **effect** of MDMA is **mood elevation**.

All Eric needed was a hat and a mobile phone to increase his serotonin level.

Drugs and Disease

Info *from the* **Human Genome Project** *is Being Used to Create* **New Drugs...**

1) The **H**uman **G**enome **P**roject (**HGP**) was a 13 year long project that **identified** all of the **genes** found in **human DNA** (the human genome).

2) The **information obtained** from the HGP is **stored** in **databases**.

3) Scientists use the databases to **identify genes**, and so **proteins**, that are **involved** in **disease**.

4) Scientists are using this information to create **new drugs** that **target** the **identified proteins**, e.g. scientists have identified an **enzyme** that **helps cancer cells** to **spread** around the body — a **drug** that **inhibits** this **enzyme** is being developed.

5) The HGP has also highlighted **common genetic variations** between people.

6) It's known that **some** of these **variations** make **some drugs less effective**, e.g. some **asthma drugs** are **less effective** for people with a **particular mutation**.

7) Drug companies can use this knowledge to design **new drugs** that **are effective** in people with these **variations**.

...but this Raises **Moral** and **Ethical Issues**

1) Creating drugs for specific genetic variations will **increase research costs** for drugs companies. These new drugs will be **more expensive**, which could lead to a **two-tier health service** — only **wealthier** people could **afford** these new drugs.

2) Some people might be **refused** an **expensive drug** because their genetic make-up indicates that it **won't be that effective** for them — it may be the **only drug available** though.

3) The **information** held within a person's genome could be **used by others**, e.g. employers or insurance companies, to **unfairly discriminate** against them. For example, if a person is **unlikely** to respond to any **drug treatments** for **cancer** an insurance company might **increase** their **life insurance premium**.

4) Revealing that a drug might not work for a person could be **psychologically damaging** to them, e.g. it could be their **only hope** to treat a disease.

Practice Questions

Q1 Name a disorder that's linked to a low level of serotonin.

Q2 Describe one way that MDMA increases the level of serotonin in the brain.

Q3 What is the Human Genome Project?

Exam Questions

Q1 Parkinson's disease affects around 120 000 people in the UK.

a) Explain the role of dopamine in controlling movement. [2 marks]

b) Describe the effect that Parkinson's disease has on the brain. [4 marks]

c) Name a drug that is used to treat Parkinson's disease and explain how it works. [4 marks]

Q2 Describe how the results of the Human Genome Project are being used to create new drugs. [5 marks]

The Minnesota Donkey and Mule Association — a different kind of MDMA...

Make sure you go back over the stuff about synaptic transmission on p. 80 — it'll really help you to understand how L-dopa and MDMA work. It's not just drugs that increase your serotonin level though, chocolate does too... which is a great excuse to gobble some down — not that you really need an excuse. I think you deserve some after all this revision.

Producing Drugs Using GMOs

Unsavoury characters in dark alleyways aren't the only things that produce drugs on demand...

Drugs *can be* Produced *Using* Genetically Modified Organisms

G̲enetically m̲odified o̲rganisms (**GMOs**) are organisms that have had their **DNA altered**. **Microorganisms**, **plants** and **animals** can all be **genetically modified** to **produce proteins** which are **used as drugs**:

1 — Genetically Modified Microorganisms

1) Here's how microorganisms are genetically engineered to produce drugs:

> 1) The **gene** for the protein (drug) is **isolated** using enzymes called **restriction enzymes**.
>
> 2) The **gene is copied** using **PCR** (see p. 35).
>
> 3) **Copies** are **inserted** into **plasmids** (small circular molecules of DNA).
>
> 4) The **plasmids** are **transferred** into **microorganisms**.
>
> 5) The **modified microorganisms** are **grown** in large containers so that they **divide** and produce **lots** of the **useful protein**, from the inserted gene.
>
> 6) The **protein** can then be **purified** and **used** as a drug.

Only drugs that are proteins can be produced by genetically modified organisms.

Plasmids are a type of vector — vectors carry genes into an organism.

2) **Lots** of drugs are produced from **genetically modified bacteria**, for example **human insulin** (used to treat **Type 1 diabetes**) and **human blood clotting factors** (used to treat **haemophilia**).

2 — Genetically Modified Plants

1) Here's how plants are genetically engineered to produce drugs:

> 1) The **gene** for the protein (drug) is **inserted** into a **bacterium** (see above).
>
> 2) The bacterium **infects** a **plant cell**.
>
> 3) The bacterium **inserts** the **gene** into the **plant cell DNA** — the **plant cell** is now **genetically modified**.
>
> 4) The **plant cell** is **grown** into an **adult plant** — the **whole plant** contains a **copy** of the **gene** in **every cell**.
>
> 5) The **protein** produced from the gene can be **purified** from the **plant tissues**, or the **protein** (drug) could be **delivered** by **eating** the **plant**.

The bacterium is used as a vector to carry the gene into the plant.

Malcolm had unwittingly eaten the Viagra plant.

2) **Some** drugs have been produced from genetically **modified** plants, for example **human insulin** and a **cholera vaccine**.

3 — Genetically Modified Animals

1) Here's how animals are genetically engineered to produce drugs:

> 1) The **gene** for the protein (drug) is **injected** into the **nucleus** of a **fertilised animal egg cell**.
>
> 2) The **egg cell** is then **implanted** into an **adult animal** — it grows into a **whole animal** that contains a **copy** of the **gene** in **every cell**.
>
> 3) The **protein** produced from the gene is normally **purified** from the **milk** of the animal.

2) Various animals have been **modified** with **human genes** to produce drugs, for example **human antithrombin** (used to treat people with a **blood clotting disorder**) has been produced from **genetically modified goats**.

Producing Drugs Using GMOs

There are **Benefits** and **Risks** Associated with Using **GMOs**

1) As well as producing drugs, GMOs are used in **agriculture** and the **food industry**.
   For example, genes for **herbicide resistance** can be inserted into **agricultural crops**.
   Herbicides can then be applied which will **kill weeds** but **not** the **herbicide-resistant crops**
   — the **genetically modified crop** will **thrive** without the weeds and this results in a **high yield**.

2) You need to know the **benefits** and **risks** associated with the **use** of **GMOs**:

Benefits

1) Agricultural **crops** can be modified so that they give **higher yields** or are **more nutritious**.
   This means these plants can be used to **reduce** the risk of **famine** and **malnutrition**.

2) Crops can also be modified to have **pest resistance**, so that **fewer pesticides** are **needed**.
   This **reduces costs** (making **food cheaper**) and **reduces** any **environmental problems**
   associated with using pesticides.

3) Industrial processes often use **enzymes**. These enzymes can be produced from genetically
   modified organisms in **large quantities** for less money, which **reduces costs**.

4) Some **disorders** can now be **treated** with **human proteins** from genetically engineered
   organisms instead of with **animal proteins**. Human proteins are **safer** and **more effective**.
   For example, **Type 1 diabetes** used to be treated with **cow insulin** but some people had an
   allergic reaction to it. **Human insulin**, produced from genetically modified **bacteria**,
   is more effective and **doesn't cause** an **allergic reaction** in humans.

5) **Vaccines** produced in plant tissues **don't need** to be **refrigerated**. This could make
   vaccines **available** to **more people**, e.g. in areas where **refrigeration** (usually needed
   for **storing** vaccines) **isn't available**.

6) **Producing drugs** using **plants** and **animals** would be **very cheap** because once the plants or
   animals are genetically modified they can be reproduced using **conventional farming methods**.
   This could make some drugs **affordable** for **more people**, especially those in poor countries.

Risks

1) Some people are **concerned** about the **transmission** of genetic material. For example,
   if **herbicide-resistant** crops **interbreed** with **wild plants** it could create 'superweeds' —
   weeds that are **resistant** to **herbicides**, and if **drug crops** interbreed with other crops
   people might end up **eating drugs they don't need** (which could be harmful).

2) Some people are worried about the **long-term impacts** of using GMOs.
   There may be **unforeseen consequences**.

3) Some people think it's **wrong** to **genetically modify animals** purely for **human benefit**.

Practice Questions

Q1 What is a genetically modified organism?

Q2 Describe how a genetically modified plant is created.

Q3 Describe how a genetically modified animal is created.

Exam Questions

Q1 Describe how the bacterium *E. coli* is genetically modified to produce human insulin. [4 marks]

Q2 Discuss the benefits and risks associated with growing a plant that has been genetically modified
to produce a hepatitis B vaccine and to be resistant to herbicides. [7 marks]

Milking a goat to get drugs — who'd have thought that was possible...

And there we go. A2 Biology. Done. Well, almost — there's still the challenge of getting all this stuff to stick in your brain. It's at times like these everyone wishes they could eat a book and absorb the information into their memory. You could try making a revision guide risotto... actually don't. Risottos can be difficult. I think I've been exposed to drugs for too long...

How to Interpret Experiment and Study Data

Science is all about getting good evidence to test your theories... so scientists need to be able to spot a badly designed experiment or study a mile off, and be able to interpret the results of an experiment or study properly. Being the cheeky little monkeys they are, your exam board will want to make sure you can do it too. Here's a quick reference section to show you how to go about interpreting data-style questions.

Here Are Some *Things* You Might be *Asked* to do...

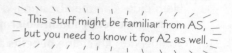
This stuff might be familiar from AS, but you need to know it for A2 as well.

Here are three examples of the kind of data you could expect to get:

Study A

An agricultural scientist investigated the effect of three different pesticides on the number of pests in wheat fields. The number of pests was estimated in each of three fields, using ground traps, before and 1 month after application of one of the pesticides. The number of pests was also estimated in a control field where no pesticide had been applied. The table shows the results.

Pesticide	Number of pests	
	Before application	1 month after application
1	89	98
2	53	11
3	172	94
Control	70	77

Study B

Study B investigated the link between the number of bees in an area and the temperature of the area. The number of bees was estimated at ten 1-acre sites. The temperature was also recorded at each site. The results are shown in the scattergram below.

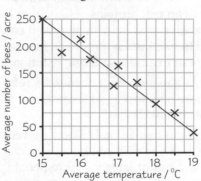

Experiment C

An experiment was conducted to investigate the effect of temperature on the rate of photosynthesis. The rate of photosynthesis in Canadian pondweed was measured at four different temperatures by measuring the volume of oxygen produced. All other variables were kept constant. The results are shown in the graph below.

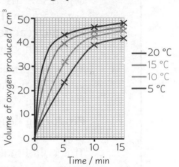

1) Describe and Manipulate the Data

You need to be able to **describe** any data you're given. The level of **detail** in your answer should be appropriate for the **number of marks** given. Loads of marks = more detail, few marks = less detail. You could also be asked to **manipulate** the data you're given (i.e. do some **calculations** on it). For the examples above:

Example — Study A

1)   You could be asked to **calculate** the **percentage change** (**increase** or **decrease**) in the number of pests for each of the pesticides and the control. E.g. for pesticide 1: (98 − 89) ÷ 89 = 0.10 = **10% increase**.

2)   You can then use these values to **describe** what the **data** shows — the **percentage increase** in pests in the field treated with **pesticide 1 was the same as for the control** (10% increase) (1 mark). **Pesticide 3 reduced** pest numbers by **45%**, but **pesticide 2** reduced the pest numbers the **most** (79% decrease) (1 mark).

Example — Study B

The data shows a **negative correlation** between the average number of bees and the temperature (1 mark).

Correlation describes the **relationship** between two variables — e.g. the one that's been changed and the one that's been measured. Data can show **three** types of correlation:

Positive   Negative   None

1)   **Positive** — as one variable **increases** the other **increases**.

2)   **Negative** — as one variable **increases** the other **decreases**.

3)   **None** — there is **no relationship** between the two variables.

Example — Experiment C

You could be asked to calculate the initial rate of photosynthesis at each temperature: The **gradient = the rate of photosynthesis**:

$$\text{Gradient} = \frac{\text{Change in Y}}{\text{Change in X}}$$

To tell if some data in a table **is correlated** — draw a **scatter diagram** of one variable against the other and **draw a line of best fit**.

How to Interpret Experiment and Study Data

2) Draw or Check a Conclusion

1) Ideally, only **two** quantities would ever change in any experiment or study — everything else would be **constant**.

2) If you can keep everything else constant and the results show a correlation then you **can** conclude that the change in one variable **does cause** the change in the other.

3) But usually all the variables **can't** be controlled, so other **factors** (that you **couldn't** keep constant) could be having an **effect**.

4) Because of this, scientists have to be very careful when **drawing conclusions**. Most results show a **link** (**correlation**) between the variables, but that **doesn't prove that a change in one causes the change in the other**.

Example — Experiment C

All other variables were **kept constant**. E.g. light intensity and CO_2 concentration **stayed the same** each time, so these **couldn't** have influenced the rate of reaction. So you **can say** that an increase in temperature up to 20 °C **causes** an increase in the rate of photosynthesis.

Example — Study B

There's a **negative correlation** between the average number of bees and temperature. But you **can't** conclude that the increase in temperature **causes** the decrease in bees. **Other factors** may have been involved, e.g. there may be **less food** in some areas, there may be **more bee predators** in some areas, or **something else** you hadn't thought of could have caused the pattern...

Example — Experiment C

A science magazine **concluded** from this data that the optimum temperature for photosynthesis is **20 °C**. The data **doesn't** support this. The rate **could** be greatest at 22 °C, or 18 °C, but you can't tell from the data because it doesn't go **higher** than 20 °C and **increases** of 5 °C at a time were used. The rates of photosynthesis at in-between temperatures **weren't** measured.

5) The **data** should always **support** the conclusion. This may sound obvious but it's easy to **jump** to conclusions. Conclusions have to be **precise** — not make sweeping generalisations.

3) Explain the Evidence

You could also be asked to **explain** the **evidence** (the data and results) — basically use your **knowledge** of the subject to explain **why** those results were obtained.

Example — Experiment C

Temperature increases the rate of photosynthesis because it **increases** the **activity** of **enzymes** involved in photosynthesis, so reactions are catalysed more quickly.

4) Comment on the Reliability of the Results

Reliable means the results can be **consistently reproduced** when an experiment or study is repeated. And if the results are reproducible they're more likely to be **true**. If the data isn't reliable for whatever reason you **can't draw** a valid **conclusion**. Here are some of the things that affect the reliability of data:

1) **Size of the data set** — For experiments, the **more repeats** you do, the **more reliable** the data. If you get the **same result** twice, it could be the correct answer. But if you get the same result **20 times**, it's much more reliable. The general rule for **studies** is the larger the **sample size**, the more **reliable** the **data** is.

E.g. Study B is quite **small** — they only studied ten 1-acre sites. The **trend** shown by the data may not appear if you studied **50 or 100 sites**, or studied them for a longer period of time.

2) **The range of values in a data set** — The **closer** all the values are to the **mean**, the **more reliable** the data set.

E.g. Study A is **repeated three more times** for pesticides 2 and 3. The percentage decrease each time is: 79%, 85%, 98% and 65% for **pesticide 2** (**mean = 82%**) and 45%, 45%, 54% and 43% for **pesticide 3** (**mean = 47%**). The data values are **closer to the mean** for pesticide 3 than pesticide 2, so that data set is **more reliable**. The **spread of values about the mean** can be shown by calculating the **standard deviation** (SD).

The **smaller the SD** the **closer** the values to the **mean** and the **more reliable the data**. SDs can be shown on a graph using **error bars**. The ends of the bars show one SD **above** and one SD **below** the **mean**.

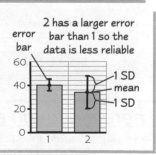

How to Interpret Experiment and Study Data

3) <u>**Variables**</u> — The **more variables** you **control**, the **more reliable** your data is. In an experiment you would control all the variables. In a study you try to control **as many as possible**.

The hat, trousers, shirt and tie variables had been well controlled in this study.

E.g. ideally, all the sites in Study B would have a similar **type** of land, similar **weather**, have the same **plants** growing, etc. Then you could be more sure that the one factor being **investigated** (temperature) is having an **effect** on the thing being **measured** (number of bees).

4) <u>**Data collection**</u> — think about all the **problems** with the **method** and see if **bias** has slipped in.

E.g. in Study A, the traps were placed on the **ground**, so pests like moths or aphids weren't included. This could have affected the results.

5) <u>**Controls**</u> — without controls, it's very difficult to **draw valid conclusions**. **Negative controls** are used to make sure that nothing you're doing in the experiment has an effect, **other than** what you're testing.

E.g. in Experiment C, the **negative control** would be all the equipment set up as normal but **without** the pondweed. If **no oxygen** was produced at any temperature it would show that the variation in the volume of oxygen produced when there was pondweed was due to the **effect** of temperature on the pondweed, and **not** the effect of temperature on **anything else** in the experiment.

6) <u>**Repetition by other scientists**</u> — for theories to become accepted as 'fact' other scientists need to **repeat** the work (see page 2). If **multiple studies** or **experiments** come to the same conclusion, then that conclusion is **more reliable**.

E.g. if a second group of scientists repeated Study B and got the same results, the results would be **more reliable**.

There Are a Few *Technical Terms* You *Need to Understand*

I'm sure you probably know these all off by heart, but it's easy to get mixed up sometimes. So here's a quick recap of some words **commonly used** when assessing and analysing experiments and studies:

1) **Variable** — A variable is a **quantity** that has the **potential to change**, e.g. weight. There are two types of variable commonly referred to in experiments:
   - **Independent variable** — the thing that's **changed** in an experiment.
   - **Dependent variable** — the thing that you **measure** in an experiment.

When drawing graphs, the dependent variable should go on the **y-axis** (the vertical axis) and the independent on the **x-axis** (the horizontal axis).

2) **Accurate** — Accurate results are those that are **really close** to the **true** answer. The true answer is **without error**, so if you can reduce error as much as possible you'll get a more accurate result. The most **accurate methods** are those that produce as **error-free** results as possible.

3) **Precise results** — These are results taken using **sensitive instruments** that measure in **small increments**, e.g. pH measured with a meter (pH 7.692) will be **more precise** than pH measured with paper (pH 8).

It's possible for results to be precise **but not** accurate, e.g. a balance that weighs to 1/1000 th of a gram will give precise results, but if it's not **calibrated** properly the results won't be accurate.

4) **Qualitative** — A **qualitative** test tells you **what's** present, e.g. an acid or an alkali.

5) **Quantitative** — A **quantitative** test tells you **how much** is present, e.g. an acid that's pH 2.46.

There's enough evidence here to conclude that data interpretation is boring...

*These pages should give you a fair idea of how to interpret data. Just use your head and remember the four things you might be asked to do — **d**escribe the **d**ata, **c**heck the **c**onclusions, **e**xplain the **e**vidence and check the **r**esults are **r**eliable.*

Answers

Unit 4: Section 1 — Photosynthesis

Page 5 — Photosynthesis and Energy Supply

1   Maximum of 6 marks available,
    from any of the 8 points below.
    In the cell, ATP is synthesised from ADP and inorganic
    phosphate/P_i [1 mark] using energy from an
    energy-releasing reaction, e.g. respiration [1 mark].
    The energy is stored as chemical energy in the phosphate
    bond [1 mark]. ATP synthase catalyses this reaction
    [1 mark]. ATP then diffuses to the part of the cell that
    needs energy [1 mark]. Here, it's broken down back
    into ADP and inorganic phosphate/P_i [1 mark], which is
    catalysed by ATPase [1 mark]. Chemical energy is
    released from the phosphate bond and used by
    the cell [1 mark].
    Make sure you don't get the two enzymes confused — ATP
    synthase **syn**thesises ATP, and ATPase breaks it down.

Page 7 — The Light-Dependent Reaction

1 a) Maximum of 1 mark available.
    The thylakoid membranes [1 mark].
  b) Maximum of 1 mark available.
    Photosystem II [1 mark].
  c) Maximum of 4 marks available.
    Light energy splits water [1 mark].
    H_2O [1 mark] $\rightarrow 2H^+ + \frac{1}{2}O_2$ [1 mark].
    The electrons from the water replace the electrons lost
    from chlorophyll [1 mark].
    The question asks you to explain the purpose of photolysis,
    so make sure you include why the water is split up —
    to replace the electrons lost from chlorophyll.
  d) Maximum of 1 mark available.
    NADP [1 mark].

Page 9 — The Light-Independent Reaction

1 a) Maximum of 6 marks available.
    Ribulose bisphosphate/RuBP and carbon dioxide/CO_2
    join together to form an unstable 6-carbon compound
    [1 mark]. This reaction is catalysed by the enzyme
    rubisco/ribulose bisphosphate carboxylase [1 mark].
    The compound breaks down into two molecules of a
    3-carbon compound called glycerate 3-phosphate/GP
    [1 mark]. Two molecules of glycerate 3-phosphate are
    then converted into two molecules of triose phosphate/
    TP [1 mark]. The energy for this reaction comes from
    ATP [1 mark] and the H^+ ions come from reduced
    NADP [1 mark].
  b) Maximum of 2 marks available.
    Ribulose bisphosphate is regenerated from triose
    phosphate/TP molecules [1 mark]. ATP provides
    the energy to do this [1 mark].
    This question is only worth two marks so only the main facts
    are needed, without the detail of the number of molecules.

c) Maximum of 3 marks available.
    No glycerate 3-phosphate/GP would be produced
    [1 mark], so no triose phosphate/TP would be produced
    [1 mark]. This means there would be no glucose
    produced [1 mark].

Unit 4: Section 2 — Ecology

Page 11 — Energy Transfer and Productivity

1 a) Maximum of 4 marks available.
    Because not all of the energy available from the grass is
    taken in by the Arctic hare [1 mark]. Some parts of the
    grass aren't eaten, so the energy isn't taken in [1 mark],
    and some parts of the grass are indigestible, so they'll
    pass through the hares and come out as waste [1 mark].
    Some energy is lost to the environment when the Arctic
    hare uses energy from respiration for things like
    movement or body heat [1 mark].
  b) Maximum of 2 marks available.
    $(137 \div 2345) \times 100 = 5.8$ [1 mark]
    Efficiency of energy transfer = 5.8% [1 mark]
    Award 2 marks for correct answer of 5.8% without
    any working.

Page 14 — Factors Affecting Abundance and Distribution

1 a) Maximum of 7 marks available.
    In the first three years, the population of prey increases
    from 5000 to 30 000. The population of predators
    increases slightly later (in the first five years), from 4000
    to 11 000 [1 mark]. This is because there's more food
    available for the predators [1 mark]. The prey population
    then falls after year three to 3000 just before year 10
    [1 mark], because lots are being eaten by the large
    population of predators [1 mark]. Shortly after the
    prey population falls, the predator population also
    falls (back to 4000 by just after year 10) [1 mark],
    because there's less food available [1 mark].
    The same pattern is repeated in years 10-20 [1 mark].
  b) Maximum of 4 marks available.
    The population of prey increased to around 40 000
    by year 26 [1 mark]. This is because there were
    fewer predators, so fewer prey were eaten [1 mark].
    The population then decreased after year 26 to 25 000
    by year 30 [1 mark]. This could be because of
    intraspecific competition [1 mark].

2 a) Maximum of 2 marks available.
    A niche is the role of a species within its habitat [1 mark].
    It includes both its biotic and abiotic interactions
    [1 mark].
  b) Maximum of 2 marks available.
    The lizards feed on the same insects, so the amount
    of food available to both species is reduced [1 mark].
    This means there will be fewer of each species in the
    area/there will be a lower abundance of each species
    than if there was only one lizard species [1 mark].

Answers

Page 17 — Investigating Populations and Abiotic Factors

1 a) Maximum of 3 marks available.
Several frame quadrats would be placed on the ground at random locations within the field *[1 mark]*. The percentage of each frame quadrat that's covered by daffodils would be recorded *[1 mark]*. The percentage cover for the whole field could then be estimated by averaging the data collected in all of the frame quadrats *[1 mark]*.

 b) Maximum of 2 marks available, from any of the 3 points below. Temperature could be measured using a thermometer *[1 mark]*. Rainfall could be measured using a rain gauge *[1 mark]*. Humidity could be measured using an electronic hygrometer *[1 mark]*.

Page 19 — Succession

1 a) Maximum of 6 marks available.
This is an example of secondary succession, because there is already a soil layer present in the field *[1 mark]*. The first species to grow will be the pioneer species, which in this case will be larger plants *[1 mark]*. These will then be replaced with shrubs and smaller trees *[1 mark]*. At each stage, different plants and animals that are better adapted for the improved conditions will move in, out-compete the species already there, and become the dominant species *[1 mark]*. As succession goes on, the ecosystem becomes more complex, so species diversity (the number and abundance of different species) increases *[1 mark]*. Eventually large trees will grow, forming the climax community, which is the final seral stage *[1 mark]*.

 b) Maximum of 2 marks available.
Ploughing destroys any plants that were growing *[1 mark]*, so larger plants may start to grow, but they won't have long enough to establish themselves before the field is ploughed again *[1 mark]*.

Unit 4: Section 3 — Global Warming

Page 21 — Introduction to Global Warming

1 Maximum of 6 marks available.
The diagram shows that the thickness of the pine tree rings fluctuated *[1 mark]*, but there was a trend of increasingly thicker rings from 1909 to 2009 *[1 mark]*. The thickness of each tree ring depends on the climate when the ring was formed *[1 mark]*. Warmer climates tend to give thicker rings than colder climates *[1 mark]*, which suggests that the climate where the pine tree lived became warmer over the last century *[1 mark]*. This is evidence for global warming *[1 mark]*.

2 Maximum of 2 marks available.
Scientists know the climate that different plant species live in now *[1 mark]*. When they find preserved pollen from similar plants, they know that the climate must have been similar when that pollen was produced *[1 mark]*.

Page 23 — Causes of Global Warming

1 a) Maximum of 4 marks available.
The temperature fluctuated between 1970 and 2008 *[1 mark]*, but the general trend was a steady increase from around 13.9 °C to around 14.4 °C *[1 mark]*. The atmospheric CO_2 concentration also showed a trend of increasing *[1 mark]* from around 328 ppm in 1970 to around 385 ppm in 2008 *[1 mark]*.
You usually have to quote figures from graphs and tables in your answer to get full marks.

 b) Maximum of 2 marks available.
There's a positive correlation between temperature and CO_2 concentration *[1 mark]*. The increasing CO_2 concentration could be linked to the increasing temperature *[1 mark]*.
You can't conclude from this data that it's a causal relationship because other factors may have been involved.

Page 25 — Effects of Global Warming

1 Maximum of 4 marks available.
An increase in temperature causes an increase in enzyme activity *[1 mark]*, which speeds up metabolic reactions *[1 mark]*. Increasing the rate of metabolic reactions in a potato tuber moth will increase its rate of growth *[1 mark]*, so it will progress through its life cycle faster *[1 mark]*.

2 Maximum of 5 marks available.
The student could plant some seedlings in soil trays, and measure the height of each seedling *[1 mark]*. The student could then put the trays in incubators at different temperatures *[1 mark]*. All other variables would need to be kept the same for each tray *[1 mark]*. After a period of incubation, the student could record the change in height of each seedling *[1 mark]*, and then calculate the average growth rate for each tray to see how increasing temperature affects growth rate *[1 mark]*.

Page 27 — Reducing Global Warming

1 Maximum of 4 marks available.
Biofuels are fuels produced from biomass / material that is or was recently living *[1 mark]*. Biofuels are burnt to release energy, which produces CO_2 *[1 mark]*. There's no net increase in atmospheric CO_2 concentration because the amount of CO_2 produced is the same as the amount of CO_2 taken in when the material was growing *[1 mark]*. Using biofuels as an alternative to fossil fuels stops the increase in atmospheric CO_2 concentration caused by burning fossil fuels *[1 mark]*.

2 Maximum of 4 marks available.
It's not actually known how emissions will change,
i.e. how accurate the extrapolated CO_2 data is *[1 mark]*.
Scientists don't know exactly how much the extrapolated
CO_2 changes will cause the global temperature to rise
[1 mark]. The change in atmospheric CO_2 concentration
due to natural causes isn't known *[1 mark]*.
Scientists also don't know what attempts there
will be to manage atmospheric CO_2 concentration,
or how successful they'll be *[1 mark]*.

Unit 4: Section 4 — Evolution

Page 29 — Evolution, Natural Selection and Speciation

1 a) Maximum of 1 mark available.
The new species could not breed with each other
[1 mark].
  b) Maximum of 3 marks available.
Different populations of flies were isolated and fed
on different foods *[1 mark]*. This caused changes in
allele frequencies between the populations *[1 mark]*,
which made them reproductively isolated and
eventually resulted in speciation *[1 mark]*.
  c) Maximum of 2 marks available, from any of the 3 points
below. Seasonal changes (become sexually active at
different times) *[1 mark]*. Mechanical changes
(changes to genitalia) *[1 mark]*. Behavioural changes
(changes in behaviour that prevent mating) *[1 mark]*.
  d) Maximum of 1 mark available, from any of the 5 points
below or any other good point.
E.g. geographical barrier *[1 mark]*, flood *[1 mark]*,
volcanic eruption *[1 mark]*, earthquake *[1 mark]*,
glacier *[1 mark]*.

Page 31 — Evidence for Evolution

1 Maximum of 5 marks available.
The theory of evolution suggests that all organisms
have evolved from shared common ancestors *[1 mark]*.
Closely related species diverged more recently *[1 mark]*.
Evolution is caused by gradual changes in the base
sequence of organisms' DNA *[1 mark]*. So, organisms
that diverged away from each other more recently
should have more similar DNA, as less time has passed
for changes in the DNA sequence to occur *[1 mark]*.
This is exactly what scientists have found *[1 mark]*.

2 Maximum of 4 marks available.
Before scientists can get their work published it must
undergo peer review, which is when other scientists who
work in that area read and review the work *[1 mark]*.
The peer reviewer checks that the work is valid and
supports the conclusion *[1 mark]*. Scientific journals
also allow other scientists to repeat experiments and
see if they get the same results using the same methods
[1 mark]. If the results are replicated over and over
again, the scientific community can be pretty confident
that the evidence collected is reliable *[1 mark]*.

Unit 4: Section 5 — Genetics

Page 34 — The Genetic Code and Protein Synthesis

1 Maximum of 5 marks available,
from any of the 6 points below.
Genes contain sections called introns that don't code for
amino acids *[1 mark]*, and sections called exons that do
code for amino acids *[1 mark]*. During transcription the
introns and exons are both copied into mRNA
[1 mark]. The introns are then removed and the exons
are joined together forming mRNA strands *[1 mark]*
by a process called splicing. The exons can be joined
together in different orders to form different mRNA
strands *[1 mark]*. This means more than one amino
acid sequence and so more than one protein can be
produced from one gene *[1 mark]*.

2 a) Maximum of 2 marks available.
$10 \times 3 = 30$ nucleotides long *[1 mark]*. Each amino acid
is coded for by three nucleotides (a codon), so the
mRNA length in nucleotides is the number of amino
acids multiplied by three *[1 mark]*.
  b) Maximum of 6 marks available.
The mRNA attaches itself to a ribosome and transfer RNA
(tRNA) molecules carry amino acids to the ribosome
[1 mark]. A tRNA molecule, with an anticodon that's
complementary to the first codon on the mRNA (the start
codon), attaches itself to the mRNA by complementary
base pairing *[1 mark]*. A second tRNA molecule
attaches itself to the next codon on the mRNA in the
same way *[1 mark]*. The two amino acids attached to
the tRNA molecules are joined by a peptide bond and
the first tRNA molecule moves away, leaving its amino
acid behind *[1 mark]*. A third tRNA molecule binds to
the next codon on the mRNA and its amino acid binds to
the first two and the second tRNA molecule moves away
[1 mark]. This process continues, producing a chain of
linked amino acids (a polypeptide chain), until there's a
stop codon on the mRNA molecule *[1 mark]*.

102

Answers

Page 37 — DNA Profiling

1 Maximum of 6 marks available.
The DNA sample is mixed with free nucleotides, primers and DNA polymerase *[1 mark]*. The mixture is heated to 95 °C to break the hydrogen bonds *[1 mark]*. The mixture is then cooled to between 50 – 65 °C to allow the primers to bind/anneal to the DNA *[1 mark]*. The primers bind/anneal to the DNA because they have a sequence that's complementary to the sequence at the start of the DNA fragment *[1 mark]*. The mixture is then heated to 72 °C and DNA polymerase lines up free nucleotides along each template strand, producing new strands of DNA *[1 mark]*. The cycle would be repeated over and over to produce lots of copies *[1 mark]*.
This question asks you to describe and explain, so you need to give the reasons why each stage is done to gain full marks.

2 a) Maximum of 5 marks available.
A sample of DNA is obtained, e.g. from a person's blood/saliva *[1 mark]*. PCR is used to amplify specific regions of the DNA *[1 mark]*. A fluorescent tag is added to all the DNA fragments *[1 mark]*. Gel electrophoresis is used to separate the DNA fragments according to their length *[1 mark]*. The separated bands produce the DNA profile, which is viewed under UV light *[1 mark]*.
b) Maximum of 2 marks available.
DNA profile 1 is most likely to be from the child's father *[1 mark]* because five out of six of the bands on his DNA profile match that of the child's, compared to only one on profile 2 *[1 mark]*.
c) Maximum of 1 mark available, from any of the 2 points below.
E.g. they can be used to link a person to a crime scene (forensic science) *[1 mark]*. Or to prevent inbreeding between animals or plants *[1 mark]*.

Unit 4: Section 6 — Microorganisms and Immunity

Page 39 — Viral and Bacterial Infections

1 Maximum of 7 marks available.
The initial symptoms of AIDS include minor infections of mucous membranes and recurring respiratory infections *[1 mark]*. These are caused by a lower than normal number of immune system cells *[1 mark]*. As AIDS progresses the number of immune system cells decreases further *[1 mark]*. Patients become susceptible to more serious infections, including chronic diarrhoea, serious bacterial infections and TB *[1 mark]*. During the late stages of AIDS, patients have a very low number of immune system cells *[1 mark]* and suffer from a range of serious infections such as toxoplasmosis and candidiasis *[1 mark]*. It's these serious infections that kill AIDS patients, not HIV itself *[1 mark]*.

2 Maximum of 3 marks available, from any of the 6 points below.
E.g. bacteria have ribosomes but viruses don't *[1 mark]*. Bacteria have a cell wall but viruses don't *[1 mark]*. Bacteria have cytoplasm, but viruses don't *[1 mark]*. Viruses have a capsid but bacteria don't *[1 mark]*. Viruses have a protein coat but bacteria don't *[1 mark]*. Viruses are smaller than bacteria *[1 mark]*.

Page 41 — Infection and The Non-Specific Immune Response

1 Maximum of 6 marks available.
Immune system cells recognise foreign antigens on the surface of a pathogen and release molecules that trigger inflammation *[1 mark]*. The molecules cause vasodilation (widening of the blood vessels) around the site of infection, increasing the blood flow to it *[1 mark]*. The molecules also increase the permeability of the blood vessels *[1 mark]*. The increased blood flow brings loads of immune system cells to the site of infection *[1 mark]* and the increased permeability allows those cells to move out of the blood vessels and into the infected tissue *[1 mark]*. The immune system cells can then start to destroy the pathogen *[1 mark]*.

2 Maximum of 6 marks available.
A phagocyte recognises the antigens on a pathogen *[1 mark]*. The phagocyte engulfs the pathogen *[1 mark]*. The pathogen is now contained in a phagocytic vacuole *[1 mark]*. A lysosome fuses with the phagocytic vacuole *[1 mark]* and digestive enzymes from the lysosome break down the pathogen *[1 mark]*. The phagocyte presents the antigens to other immune system cells *[1 mark]*.

Page 43 — The Specific Immune Response

1 a) Maximum of 2 marks available.
T cells are activated when receptors on the surface of the T cells *[1 mark]* bind to complementary antigens presented to them by phagocytes *[1 mark]*.
b) Maximum of 3 marks available.
T helper cells *[1 mark]*, T killer cells *[1 mark]* and T memory cells *[1 mark]*.
2 Maximum of 3 marks available.
Antibodies agglutinate pathogens, so that phagocytes can get rid of a lot of the pathogens at once *[1 mark]*. Antibodies neutralise toxins produced by pathogens *[1 mark]*. Antibodies bind to pathogens to prevent them from binding to and infecting human cells *[1 mark]*.
There are three marks available for this question so you need to think of three different functions.

Answers

Page 45 — Developing Immunity

1   Maximum of 10 marks available.
    Before the person is exposed to the antigen there
    are none of the right antibodies in their bloodstream
    [1 mark]. This is because the T and B cells haven't come
    into contact with the antigen *[1 mark]*. Shortly after the
    person is exposed to the antigen the concentration of the
    right antibody begins to rise *[1 mark]*. This is the primary
    response and it's slow because there aren't many B cells
    that can make the antibody that binds to the antigen
    [1 mark]. The concentration of antibody peaks at about
    20 days after the first exposure and then it begins to fall
    [1 mark]. It begins to fall because the person's immune
    system is starting to overcome the infection *[1 mark]*.
    When the person is exposed to the same pathogen again
    at 60 days, the secondary response happens *[1 mark]*.
    The concentration of antibody rises quickly from the
    moment of the second exposure to a peak at about 7
    days after the second exposure *[1 mark]*. This is because
    memory B cells that were created after the first exposure
    quickly divide into plasma cells *[1 mark]*. They produce
    the right antibody to the antigen almost immediately
    [1 mark].

2   Maximum of 6 marks available.
    HIV kills the immune systems cells that it infects
    [1 mark]. This reduces the overall number of immune
    system cells in the body, which reduces the chance of
    HIV being detected *[1 mark]*. HIV has a high rate of
    mutation in the genes that code for antigen proteins.
    The mutations change the structure of the antigens
    forming new strains of the virus. This is called antigenic
    variation *[1 mark]*. It means the antibodies produced
    for one strain of HIV won't recognise other strains with
    different antigens, so the immune system has to produce
    a primary response against each new strain *[1 mark]*.
    HIV disrupts antigen presentation in infected cells
    [1 mark]. This prevents immune system cells
    recognising and killing the infected cells *[1 mark]*.

Page 47 — Antibiotics

1 a) Maximum of 4 marks available.
    The bacteria to be tested are spread onto the agar plate
    [1 mark]. Paper discs soaked with the antibiotics are
    placed apart on the plate, along with a negative control
    disc soaked in sterile water *[1 mark]*. The whole
    experiment is performed using aseptic techniques, e.g.
    using Bunsen burners to sterilise instruments *[1 mark]*.
    The plate is incubated at 25-30 °C for 24-36 hours
    [1 mark].
  b) Maximum of 2 marks available.
    Erythromycin *[1 mark]*, because it has the largest
    inhibition zone *[1 mark]*.

2   Maximum of 2 marks available, 1 mark for a description
    of poor hygiene and 1 mark for a code of practice.
    E.g. Hospital staff and visitors not washing their hands
    before and after visiting a patient *[1 mark]*. Hospital staff
    and visitors should be encouraged to wash their hands
    before and after they've been with a patient *[1 mark]*.
    Equipment (e.g. beds or surgical instruments) and
    surfaces not being disinfected after they're used *[1 mark]*.
    Equipment and surfaces should be disinfected after
    they're used *[1 mark]*.

Page 49 — Microbial Decomposition and Time of Death

1   Maximum of 8 marks available.
    A dead human body loses heat at a rate of approximately
    1.5-2.0 °C per hour *[1 mark]*, which suggests the time
    of death was around 4-5 hours ago/at 17:45-18:45
    [1 mark]. Rigor mortis has only recently started as it's
    limited to the upper parts of the body *[1 mark]*, which
    suggests that the time of death was around 4 to 6 hours
    ago *[1 mark]*. There's a lack of visible decomposition
    [1 mark], which suggests that the time of death was only
    a few hours ago *[1 mark]*. Blowfly larvae hatch from
    eggs approximately 24 hours after being laid *[1 mark]*,
    so no blowfly larvae on the body suggests that the time
    of death was less than 24 hours ago *[1 mark]*.

Unit 5: Section 1 — Muscles and Respiration

Page 51 — Muscles and Movement

1   Maximum of 3 marks available.
    Muscles are made up of bundles of muscle fibres
    [1 mark]. Muscle fibres contain long organelles called
    myofibrils *[1 mark]*. Myofibrils contain bundles of
    myofilaments *[1 mark]*.
2 a) Maximum of 2 marks available.
    The quadriceps are the extensors *[1 mark]*.
    The hamstrings are the flexors *[1 mark]*.
  b) Maximum of 1 mark available.
    Antagonistic pairs *[1 mark]*.

3   Maximum of 3 marks available.
    A = sarcomere *[1 mark]*.
    B = Z-line *[1 mark]*.
    C = H-zone *[1 mark]*.

Page 53 — Muscle Contraction

1   Maximum of 3 marks available.
    Muscles need ATP to relax because ATP provides the
    energy to break the actin-myosin cross bridges *[1 mark]*.
    If the cross bridges can't be broken, the myosin heads
    will remain attached to the actin filaments *[1 mark]*,
    so the actin filaments can't slide back to their relaxed
    position *[1 mark]*.

Answers

2   Maximum of 3 marks available.
     The muscles won't contract *[1 mark]* because calcium
     ions won't be released into the sarcoplasm, so troponin
     won't be removed from its binding site *[1 mark]*.
     This means no actin-myosin cross bridges
     can be formed *[1 mark]*.

Page 55 — Aerobic Respiration

1 a) Maximum of 6 marks available,
     from any of the 7 points below.
     First, the 6-carbon glucose molecule is phosphorylated
     [1 mark] by adding two phosphates from two molecules
     of ATP *[1 mark]*. This creates two molecules of triose
     phosphate *[1 mark]* and two molecules of ADP *[1 mark]*.
     Triose phosphate is oxidised (by removing hydrogen)
     to give two molecules of 3-carbon pyruvate *[1 mark]*.
     The hydrogen is accepted by two molecules of NAD,
     producing two molecules of reduced NAD *[1 mark]*.
     During oxidation four molecules of ATP are produced
     [1 mark].
     When describing glycolysis make sure you get the number of
     molecules correct — one glucose molecule produces two
     molecules of triose phosphate. You could draw a diagram
     in the exam to show the reactions.
  b) Maximum of 1 mark available.
     2 ATP molecules *[1 mark]*

2   Maximum of 3 marks available,
     from any of the 4 points below.
     The 3-carbon pyruvate is decarboxylated *[1 mark]*, then
     converted to acetate by the reduction of NAD *[1 mark]*.
     Acetate combines with coenzyme A (CoA) to
     form acetyl coenzyme A (acetyl CoA) *[1 mark]*.
     No ATP is produced *[1 mark]*.

Page 57 — Aerobic Respiration

1 a) Maximum of 2 mark available.
     The transfer of electrons down the electron transport
     chain stops *[1 mark]*. So there's no energy released to
     phosphorylate ADP/produce ATP *[1 mark]*.
  b) Maximum of 2 marks available.
     The Krebs cycle stops *[1 mark]* because there's
     no oxidised NAD/FAD coming from the
     electron transport chain *[1 mark]*.
     Part b is a bit tricky — remember that when the electron
     transport chain is inhibited, the reactions that depend
     on the products of the chain are also affected.

Page 59 — Respirometers and
Anaerobic Respiration

1 a) Maximum of 1 mark available.
     To make sure the results are only due to oxygen
     uptake by the woodlouse *[1 mark]*.

  b) Maximum of 2 marks available.
     The oxygen taken up would be replaced by carbon
     dioxide given out / there would be no change in air
     volume in the test tube *[1 mark]*. This means there
     would be no movement of the liquid in the manometer
     [1 mark].
  c) Maximum of 1 mark available.
     Carbon dioxide/CO_2 *[1 mark]*.

2   Maximum of 1 mark available.
     Because lactate fermentation doesn't involve electron
     carriers/the electron transport chain/oxidative
     phosphorylation *[1 mark]*.

Unit 5: Section 2 — Exercise

Page 61 — Electrical Activity in the Heart

1 a) Maximum of 1 mark available.
     The sinoatrial node acts as a pacemaker /
     initiates heartbeats *[1 mark]*.
  b) Maximum of 1 mark available.
     The bundle of His conducts the waves of electrical
     activity from the AVN to the Purkyne fibres in the
     ventricle walls *[1 mark]*.

2   Maximum of 2 marks available.
     The ventricle is not contracting properly *[1 mark]*.
     This could be because of muscle damage / because the
     AVN is not conducting impulses to the ventricles
     properly *[1 mark]*.

Page 63 — Variations in Heart Rate
and Breathing Rate

1 a) Maximum of 5 marks available.
     The breathing rate would go up *[1 mark]*, because
     carbonic acid lowers blood pH *[1 mark]*. This stimulates
     chemoreceptors in the medulla, aortic bodies and carotid
     bodies *[1 mark]*. The chemoreceptors send nerve
     impulses to the medulla *[1 mark]*. In turn, the medulla
     sends more frequent nerve impulses to the intercostal
     muscles and diaphragm *[1 mark]*.
     This question only asks about the breathing rate, so you
     won't get any extra marks for commenting on the depth of
     breathing or speed of gas exchange.
  b) Maximum of 1 mark available.
     cardiac output (cm³/min) = heart rate (beats per minute)
     × stroke volume (cm³) *[1 mark]*.

Page 65 — Investigating Ventilation

1 a) Maximum of 2 marks available.
     Tidal volume = 1.4 − 1.0 = 0.4 dm³ *[1 mark]*.
     Breathing rate = 12 breaths per minute/bpm *[1 mark]*.
  b) Maximum of 2 marks available.
     The tidal volume would be larger *[1 mark]* and the
     breathing rate would be faster *[1 mark]*.

Answers

Page 67 — Homeostasis

1   Maximum of 3 marks available.
Receptors detect when a level is too high or too low
[1 mark], and the information's communicated via the
nervous system or the hormonal system to effectors
[1 mark]. Effectors respond to counteract the change /
to bring the level back to normal *[1 mark]*.

2 a) Maximum of 3 marks available.
Thermoreceptors/temperature receptors detect a
higher internal temperature than normal due to exercise
[1 mark]. The thermoreceptors/temperature receptors
send impulses along sensory neurones to the
hypothalamus *[1 mark]*. The hypothalamus sends
impulses along motor neurones to effectors (e.g. sweat
glands) to reduce body temperature *[1 mark]*.
  b) Maximum of 2 marks available,
from any of the 3 points below.
Sweating *[1 mark]*, hairs lying flat *[1 mark]*
and vasodilation of arterioles *[1 mark]*.

Page 69 — Exercise and Health

1 a) Maximum of 1 mark available.
The table shows that the relative risk of CHD in women
increases with less physical activity *[1 mark]*.
  b) Maximum of 1 mark available.
There's a correlation/link between lower levels of
physical activity in women and an increased risk of CHD
[1 mark].

Page 71 — Exercise and Health

1 a) Maximum of 3 marks available,
from any of the 6 points below.
Keyhole surgery involves a much smaller incision than
open surgery so the patient loses less blood *[1 mark]* and
has less scarring of the skin *[1 mark]*. The patient usually
suffers less pain after keyhole surgery than open surgery
[1 mark] and the patient usually recovers more quickly
[1 mark]. It's usually easier for the patient to return to
normal activities after keyhole surgery than open surgery
[1 mark] and their hospital stay is usually shorter
[1 mark].
  b) Maximum of 2 marks available.
A prosthesis could be used to replace his knee *[1 mark]*.
This might make it possible for him to play sport again
[1 mark].

2   Maximum of 2 marks available.
Banning the use of performance-enhancing drugs makes
competitions fairer *[1 mark]*. Athletes are less tempted to
take drugs that can have serious health risks *[1 mark]*.

Unit 5: Section 3 — Responding to the Environment

Page 73 — Nervous and Hormonal Communication

1   Maximum of 5 marks available.
Receptors detect the stimulus *[1 mark]*, e.g. light
receptors/photoreceptors in the animal's eyes detect
the bright light *[1 mark]*. The receptors send impulses
along neurones via the CNS to the effectors *[1 mark]*.
The effectors bring about a response *[1 mark]*,
e.g. the circular iris muscles contract to constrict
the pupils and protect the eyes *[1 mark]*.

2   Maximum of 3 marks available,
from any of the 4 points below.
The nervous system sends information as electrical
impulses but the hormonal system sends information as
chemicals *[1 mark]*. Nervous responses are faster than
hormonal responses *[1 mark]*. Nervous responses are
localised but hormonal responses are widespread
[1 mark]. Nervous responses are short-lived but
hormonal responses are long-lasting *[1 mark]*.

Page 75 — The Nervous System — Receptors

1   Maximum of 7 marks available.
Light energy bleaches rhodopsin / causes rhodopsin to
break apart into retinal and opsin *[1 mark]*. This causes
the sodium ion channels to close *[1 mark]*. So sodium
ions are still actively transported out of the cell but they
can't diffuse back in *[1 mark]*. This means sodium ions
build up on the outside of the cell, making the cell
membrane hyperpolarised *[1 mark]*. This causes the
rod cell to stop releasing neurotransmitters *[1 mark]*.
There's no inhibition of the bipolar neurone *[1 mark]*,
so the bipolar neurone depolarises and sends an action
potential to the brain via the optic nerve *[1 mark]*.

Page 77 — The Nervous System — Neurones

1 a) Maximum of 1 mark available.
Stimulus *[1 mark]*.
  b) Maximum of 3 marks available.
A stimulus causes sodium ion channels in the neurone
cell membrane to open *[1 mark]*. Sodium ions diffuse
into the cell *[1 mark]*, so the membrane becomes
depolarised *[1 mark]*.
  c) Maximum of 2 marks available.
The membrane was in the refractory period *[1 mark]*,
so the sodium ion channels were recovering
and couldn't be opened *[1 mark]*.

Answers

Page 79 — The Nervous System — Neurones

1 Maximum of 5 marks available.
Transmission of action potentials will be slower in neurones with damaged myelin sheaths *[1 mark]*. This is because myelin is an electrical insulator *[1 mark]*, so increases the speed of action potential conduction *[1 mark]*. The action potentials 'jump' between the nodes of Ranvier/between the myelin sheaths *[1 mark]*, where sodium ion channels are concentrated *[1 mark]*.
Don't panic if a question mentions something you haven't learnt about. You might not know anything about multiple sclerosis but that's fine, because you're not supposed to. All you need to know to get full marks here is how myelination affects the speed of action potential conduction.

Page 81 — The Nervous System — Synapses

1 Maximum of 5 marks available.
A — presynaptic membrane *[1 mark]*.
B — vesicle/vesicle containing neurotransmitter *[1 mark]*.
C — synaptic cleft *[1 mark]*.
D — postsynaptic receptor *[1 mark]*.
E — postsynaptic membrane *[1 mark]*.

2 Maximum of 6 marks available,
from any of the 10 points below.
The action potential arriving at the presynaptic membrane stimulates voltage-gated calcium ion channels to open *[1 mark]*, so calcium ions diffuse into the neurone *[1 mark]*. This causes synaptic vesicles, containing neurotransmitter, to move to the presynaptic membrane *[1 mark]*. They then fuse with the presynaptic membrane *[1 mark]*. The vesicles release the neurotransmitter into the synaptic cleft *[1 mark]*. The neurotransmitter diffuses across the synaptic cleft *[1 mark]* and binds to specific receptors on the postsynaptic membrane *[1 mark]*. This causes sodium ion channels in the postsynaptic membrane to open *[1 mark]*. The influx of sodium ions causes depolarisation *[1 mark]*. This triggers a new action potential to be generated at the postsynaptic membrane *[1 mark]*.

Page 83 — Responses in Plants

1 Maximum of 3 marks available.
IAA is produced in the tip of shoots and is moved around the plant, so different parts of the plant have different amounts of IAA *[1 mark]*. The uneven distribution of IAA means there's uneven growth of the plant *[1 mark]*. IAA moves to the more shaded parts of the shoots, making the cells there elongate, which makes the shoot bend towards the light *[1 mark]*.

2 a) Maximum of 1 mark available.
P_{FR} is a phytochrome molecule in a state that absorbs far-red light/light at a wavelength of 730 nm *[1 mark]*.
b) Maximum of 4 marks available.
An iris would flower in the summer/June to August *[1 mark]* because it's stimulated to flower by high levels of P_{FR}, which occurs when nights are short *[1 mark]*. Daylight contains more red light than far-red light, so more P_R is converted into P_{FR} than P_{FR} is converted into P_R *[1 mark]*. When nights are short in the summer, there's not much time for P_{FR} to be converted back into P_R, so P_{FR} builds up *[1 mark]*.

Unit 5: Section 4 — The Brain and Behaviour

Page 86 — Brain Structure and Function

1 a) Maximum of 1 mark available.
Hypothalamus *[1 mark]*.
b) Maximum of 2 marks available.
Control of breathing *[1 mark]*.
Control of heart rate *[1 mark]*.
c) Maximum of 1 mark available.
Lack of coordinated movement / balance *[1 mark]*.
You know that the cerebellum normally coordinates movement, so damage to it is likely to cause a lack of coordinated movement or balance.

2 a) Maximum of 2 marks available,
from any of the 3 points below.
An MRI scan would give information about the extent of the bleeding *[1 mark]*, the location of the bleeding *[1 mark]* and what brain functions might be affected by the bleeding *[1 mark]*.
b) Maximum of 1 mark available.
Functional magnetic resonance imaging/fMRI *[1 mark]*.

Page 89 — Brain Development and Habituation

1 a) Maximum of 1 mark available.
'Nature' means your genes *[1 mark]*.
b) Maximum of 2 marks available.
A newborn baby's brain hasn't really been affected by the environment *[1 mark]*. This means scientists can see which aspects of brain development are more likely to be due to nature than nurture *[1 mark]*.
c) Maximum of 2 marks available,
from any of the 3 points below.
Twin studies *[1 mark]*. Brain damage studies *[1 mark]*. Cross-cultural studies *[1 mark]*.
The question asks you to suggest two types of study to directly investigate the effect of nature and nurture on brain development in <u>humans</u>, so don't go writing about animal experiments.

Answers

2 a) Maximum of 3 marks available.
The birds' behaviour is habituation because they showed a reduced response (they didn't fly away as much) **[1 mark]** to the unimportant stimulus of the birdwatcher **[1 mark]** after repeated exposure for an hour every day **[1 mark]**.

b) Maximum of 1 mark available.
Habituation means the birds don't waste time and energy responding to unimportant stimuli **[1 mark]**.

Page 91 — Development of the Visual Cortex

1 a) Maximum of 4 marks available.
Hubel and Wiesel stitched shut one eye of very young kittens for several months **[1 mark]**. When they unstitched the eyes, Hubel and Wiesel found that the kitten's eye that had been stitched up was blind **[1 mark]**. They also found the ocular dominance columns that were stimulated by the open eye had become bigger and had taken over the ocular dominance columns that weren't visually stimulated/for the shut eye **[1 mark]**. Hubel and Wiesel's experiments showed that the visual cortex only develops properly if both eyes are visually stimulated in the very early stages of life **[1 mark]**.

b) Maximum of 2 marks available.
Yes, their experiments give evidence for a critical 'window' in the development of the human visual system because our visual cortex is also made up of ocular dominance columns **[1 mark]**. The critical 'window' is the period of time in very early life when it's critical that you're exposed to the right visual stimuli for the visual system to develop properly **[1 mark]**.

c) Maximum of 2 marks available, from any of the 4 points below.
Animal research has led to lots of medical breakthroughs, e.g. antibiotics **[1 mark]**. Animal experiments are only done when necessary and scientists follow strict rules **[1 mark]**. Using animals is currently the only way to study how a drug affects the whole body **[1 mark]**. Some people think humans have a greater right to life than animals **[1 mark]**.

Unit 5: Section 5 — Drugs

Page 93 — Drugs and Disease

1 a) Maximum of 2 marks available.
Dopamine transmits nerve impulses across synapses **[1 mark]** in the parts of the brain that control movement **[1 mark]**.

b) Maximum of 4 marks available.
In Parkinson's disease the neurones in the parts of the brain that control movement are destroyed **[1 mark]**. These neurones normally produce the neurotransmitter dopamine **[1 mark]**, so losing them causes a lack of dopamine **[1 mark]**. This causes a decrease in the transmission of the nerve impulses involved in movement **[1 mark]**.

c) Maximum of 4 marks available.
L-dopa is a drug that's used to treat Parkinson's disease **[1 mark]**. L-dopa is absorbed into the brain and converted into dopamine by the enzyme dopa-decarboxylase **[1 mark]**. This increases the level of dopamine in the brain **[1 mark]**, which causes an increase in the transmission of the nerve impulses involved in movement **[1 mark]**.

2 Maximum of 5 marks available.
Scientists use databases that store the information from the HGP to identify proteins that are involved in disease **[1 mark]**. Scientists are using this information to create new drugs that target the identified proteins **[1 mark]**. The HGP has also highlighted common genetic variations between people **[1 mark]**. It's known that some of these variations make some drugs less effective **[1 mark]**. Drug companies are using this knowledge to design new drugs that are effective in people with these variations **[1 mark]**.

Page 95 — Producing Drugs Using GMOs

1 Maximum of 4 marks available.
The human insulin gene is isolated using enzymes called restriction enzymes **[1 mark]**. The gene is then copied using PCR and the copies of the gene are inserted into plasmids **[1 mark]**. The plasmids are transferred into microorganisms **[1 mark]**. The modified microorganisms are grown so that they divide and produce lots of human insulin **[1 mark]**.

2 Maximum of 7 marks available.
The hepatitis B vaccine in the plant tissues won't need to be refrigerated **[1 mark]**. This could make the vaccine available to people in areas where refrigeration isn't available **[1 mark]**. Herbicide resistance is a benefit because the plants will be unaffected by herbicides **[1 mark]**. The genetically modified plant will thrive after weeds are killed by herbicides and this will give a high yield of the vaccine **[1 mark]**. However, the transmission of genetic material between the genetically modified plants and wild plants could occur **[1 mark]**. This could create superweeds that are resistant to herbicides **[1 mark]**. There may also be unforeseen consequences from using the genetically modified plant **[1 mark]**.
The question asked you to discuss the benefits and risks of growing the plant — make sure you write about both.

Index

Index

Index